THE FINITE ELEMENT METHOD AND ITS APPLICATIONS

THE FINITE ELEMENT METHOD AND ITS APPLICATIONS

MASATAKE MORI

MACMILLAN PUBLISHING COMPANY
NEW YORK

COLLIER MACMILLAN CANADA, INC.
TORONTO

COLLIER MACMILLAN PUBLISHERS
LONDON

Macmillan Publishing Company
866 Third Avenue, New York, NY 10022

Collier Macmillan Canada, Inc.

Printed in the United States of America

Printing: 1 2 3 4 5 6 7 8 9 10   Year: 6 7 8 9 0 1 2 3 4 5

Library of Congress Cataloging-in-Publication Data

Mori, Masatake.
  The finite element method and its applications.

  Translation of: Yūgen yōsohō to sono ōyō.
  Bibliography: p.
  Includes index.
  1. Finite element method. I. Title.
TA347.F5M6713 1986 620'.0015153'53   86-8756
ISBN 0-02-948621-1

Contents

Preface to English Edition

This is a translation of *Yuugen-youso-hou To Sono Ouyou,* written in Japanese. The translation was done by the author of the original edition. Since the original edition was intended to present the fundamentals of the finite element method, and to avoid including material that might be out of date in several years, the original edition has been translated faithfully into English without changing the contents, except that books [7], [8], and [10] on the reference list, which were written in Japanese, have been replaced by equivalent books written in English.

I have attempted to correct errors in the original edition and wish to thank those who have pointed out such errors, in particular, Shinsuke Suga. Thanks are also due to Atsuko Yamamoto and Tami Nakata for typing the manuscript.

Masatake Mori

Preface

The purpose of this book is to introduce the finite element method from the standpoint of applied mathematics. While this technique is indispensable in the field of structural mechanics, it is also a powerful tool for the numerical solution of partial differential equations related to various natural phenomena. Actually it has been a long time since the finite element method established itself in the natural sciences and in engineering. Accordingly, a variety of books about this method have been published. There are two conventional ways to approach the finite element method—from structural mechanics and from the solution of partial differential equations. Consequently, there are also two standpoints when writing a book on this subject. I intended to write this book from the standpoint of solving partial differential equations. Furthermore, books on the finite element method may be classified into two groups—books about techniques and books about mathematics. This book belongs in the latter group.

Although in practice the finite element method is applied to problems in two or three space dimensions, this book starts with applications to problems in one dimension in order to provide an easily understood description of the basic idea of the method, and then presents the mathematical background and error analysis. The main part of the book describes the application of the finite element method to partial differential equations in two dimensions, including time-dependent equations such as the heat equation and the wave equation. Chapters 3, 6, 7, and 14 are devoted mainly to error analysis, and the description is more mathematical, so that these chapters may be omitted on a first reading if the reader is more interested in the techniques of the finite element method.

In practice, in the final stage of the finite element method a system of linear equations with a large, sparse coefficient matrix of order, say, several tens of thousands must be solved, or an eigenvalue problem with respect to such a matrix must be solved. Therefore, in order to master thoroughly the techniques of the finite element method numerical analysis of such large-scale matrices should be learned. However, the numerical analysis of linear algebra is itself a large problem, and to describe it is beyond the scope of this book, hence only brief comments on this subject are included.

The amount of work published in the field of the finite element method is enormous, and it is impossible to describe all of it in a single book. Therefore the material presented is not exhaustive but is limited to what is necessary for readers studying this procedure for the first time. In particular, in Chap. 13, methods but not theories for solving nonlinear problems are presented. I hope that those who become interested in this method after reading the book will expand their knowledge by consulting the references listed at the end.

I intended to write this book so that it could be read without referring to other books or papers. In order to learn about the finite element method knowledge of the variational principle is necessary. And in order to understand it mathematically a knowledge of functional analysis is required. However, this book does not assume this specialized knowledge. On the contrary, the reader will be able to learn about the variational principle and functional analysis in a practical way using the finite element method. Although this book belongs to the field of mathematics, I have avoided descriptions that are too abstract and tried to write so that the reader can understand the mathematical theorems intuitively. In this way, I hoped to bridge the gap between the theory and the practice of the finite element method. Furthermore, in the theoretical discussions I have tried to explain modern ideas using classical terms so that physicists and engineers who are not familiar with the technical terms of modern mathematics can read the book without difficulty. I hope that this book will be useful not only as an introductory text for students studying the finite element method for the first time but also in satisfying the mathematical interests of scientists and engineers who have used the finite element method as a tool for solving problems.

Special thanks are due to Professor Hiroshi Hujita who aroused my interest in the mathematical aspect of the finite element method. I also thank my many colleagues, especially Makoto Natori, Masaaki Nakamura, and Masaaki Sugihara, who read the manuscript carefully and provided many suggestions. Finally, thanks are due to Mr. Hisao Miyauchi of Iwanami Shoten.

Masatake Mori

THE FINITE ELEMENT METHOD AND ITS APPLICATIONS

1
___

The Basic Idea of the Finite Element Method

1.1 *A Two-Point Boundary Value Problem*

In order to sketch the basic idea of the *finite element method* (FEM) we first consider a two-point boundary value problem in one dimension:

$$-\frac{d}{dx}\left(p\frac{du}{dx}\right) + qu = f(x) \qquad 0 < x < 1 \tag{1.1.1}$$

$$u(0) = u(1) = 0 \tag{1.1.2}$$

where p and q are given positive constants and $f(x)$ is a given function. The boundary condition (1.1.2), which prescribes the values at both end points to vanish, is a typical example of a homogeneous Dirichlet condition.

Since (1.1.1) is a linear differential equation with constant coefficients having an inhomogeneous term, it is easy to solve for u in a closed form. However, we consider here an approximate solution in terms of a Fourier series with N terms as follows in order to explain the basic idea of the FEM:

$$u_N(x) = \sum_{j=1}^{N} a_j \sin j\pi x \tag{1.1.3}$$

The terms of $\cos j\pi x$ that may appear in a general Fourier series are discarded at the beginning in the present series so that it satisfies the boundary condition (1.1.2). In order to obtain the values of the coefficients a_j we

1

use a conventional procedure; that is, we substitute u_N in (1.1.3) for u in (1.1.1), multiply both sides by $\sin k\pi x$, and integrate over $(0,1)$. Then we have

$$\sum_{j=1}^{N} a_j p (j\pi)^2 \int_0^1 \sin k\pi x \, \sin j\pi x \, dx + \sum_{j=1}^{N} a_j q \int_0^1 \sin k\pi x \, \sin j\pi x \, dx$$

$$= \int_0^1 f(x) \, \sin k\pi x \, dx \qquad (1.1.4)$$

From the orthogonality of $\{\sin j\pi x\}$ over $(0,1)$,

$$\int_0^1 \sin k\pi x \, \sin j\pi x \, dx = \begin{cases} \frac{1}{2} & k = j \\ 0 & k \neq j \end{cases} \qquad (1.1.5)$$

we have

$$a_k = \frac{2}{(k\pi)^2 p + q} \int_0^1 f(x) \, \sin k\pi x \, dx \qquad (1.1.6)$$

Therefore if we substitute this expression into (1.1.3) we eventually obtain an approximate solution to the problem (1.1.1) and (1.1.2).

1.2 A Solution in Terms of a Generalized Fourier Series

We can generalize the procedure stated above in the following way. First we choose a set of linearly independent functions

$$\varphi_j(x) \qquad j = 1, 2, \dots, N \qquad (1.2.1)$$

and construct an approximate solution to the boundary value problem (1.1.1) and (1.1.2) in terms of a linear combination of these functions:

$$u_N(x) = \sum_{j=1}^{N} a_j \varphi_j(x) \qquad (1.2.2)$$

We assume here that each $\varphi_j(x)$ satisfies

$$\varphi_j(0) = \varphi_j(1) = 0 \qquad (1.2.3)$$

This assumption forces $u_N(x)$ to satisfy the boundary condition

$$u_N(0) = u_N(1) = 0 \qquad (1.2.4)$$

corresponding to (1.1.2) from the beginning. Substituting $u_N(x)$ for u in (1.1.1), multiplying both sides by $\phi_k(x)$, and integrating over $(0,1)$, we have

$$- \sum_{j=1}^{N} a_j p \int_0^1 \varphi_k(x) \frac{d^2 \varphi_j(x)}{dx^2} \, dx + \sum_{j=1}^{N} a_j q \int_0^1 \varphi_k(x) \varphi_j(x) \, dx$$

$$= \int_0^1 f(x)\varphi_k(x) \, dx \qquad (1.2.5)$$

Integrating the first term by parts using the boundary condition (1.2.3) results in

$$\sum_{j=1}^N a_j \left\{ p \int_0^1 \frac{d\varphi_k}{dx} \frac{d\varphi_j}{dx} \, dx + q \int_0^1 \varphi_k \varphi_j \, dx \right\} = \int_0^1 f(x)\varphi_k(x) \, dx$$

$$k = 1, 2, \ldots, N \qquad (1.2.6)$$

The set of functions

$$\varphi_j(x) = \sin j\pi x \qquad j = 1, 2, \ldots, N \qquad (1.2.7)$$

mentioned above has an orthogonality in the sense of (1.1.5) over (0,1). In addition, their derivatives

$$\frac{d\varphi_j(x)}{dx} = j\pi \cos j\pi x \qquad j = 1, 2, \ldots, N \qquad (1.2.8)$$

also have the same type of orthogonality. Therefore only the term with $j = k$ on the left side of (1.2.6) remains without vanishing, and the coefficients $\{a_j\}$ can be obtained by simple algebraic division. An arbitrarily given set of functions $\{\phi_j(x)\}$, on the other hand, usually has no orthogonality. Even if these functions $\{\phi_j(x)\}$ satisfy an orthogonality requirement by themselves, their derivatives $\{d\phi_j(x)/dx\}$ in general will not. Also, if p or q is a function of x, the orthogonality of the functions $\{\phi_j(x)\}$ will gain nothing at all.

The system of equations (1.2.6), which is obtained on the basis of $\{\phi_j(x)\}$, is a system of simultaneous linear equations with N unknowns $\{a_j\}$. If the coefficient matrix of (1.2.6) is not singular, we can obtain a solution by solving this system of equations for a_j, $j = 1, 2, \ldots, N$. Each member of the set of functions given by (1.2.1) is called a *basis function*, and the method for obtaining a solution in the form of a Fourier series in a wider sense as shown above is called *Galerkin's method*. Typical basis functions that have been conventionally used in applied analysis are trigonometric functions, simple monomials, and orthogonal polynomials.

1.3 Piecewise Linear Basis Functions

Consider a pyramid-shaped set of basis functions $\phi_k(x)$ as shown in Fig. 1.1. In order to define this set of functions we first divide the interval [0,1] into n subintervals with an equal mesh size

$$h = \frac{1}{n} \qquad (1.3.1)$$

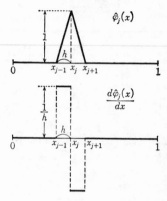

Fig. 1.1 Piecewise linear basis function and its derivative.

Then, in each subinterval bounded by the *nodes*

$$x_k = kh \qquad k = 0, 1, 2,..., n \qquad (1.3.2)$$

we define

$$\varphi_k(x) = \begin{cases} 0 & 0 \le x < x_{k-1} \\ \dfrac{x - x_{k-1}}{h} & x_{k-1} \le x < x_k \\ \dfrac{x_{k+1} - x}{h} & x_k \le x < x_{k+1} \\ 0 & x_{k+1} \le x \le 1 \end{cases} \qquad (1.3.3)$$

where $\hat{\phi}_0(x)$ and $\hat{\phi}_n(x)$ are defined by the right half and left half of (1.3.3), respectively. This type of basis function is called a *piecewise linear basis function*. One of the remarkable characteristics of this function is that its support is extremely localized to a small domain. Such a function is sometimes called a *local basis function*. The fact that the support of the basis function is localized is, as will be shown later, significant from the standpoint of numerical computation.

The derivatives of the basis function (1.3.3) are

$$\frac{d\varphi_k}{dx} = \begin{cases} 0 & 0 \le x < x_{k-1} \\ \dfrac{1}{h} & x_{k-1} \le x < x_k \\ -\dfrac{1}{h} & x_k \le x < x_{k+1} \\ 0 & x_{k+1} \le x \le 1 \end{cases} \qquad (1.3.4)$$

as shown in Fig. 1.1.

1.4 *Construction of an Approximate Equation*

We write here an approximate solution of (1.1.1) and (1.1.2) in terms of a linear combination of $\{\phi_j(x)\}$:

$$\hat{u}_n(x) = \sum_{j=1}^{n-1} a_j \hat{\varphi}_j(x) \qquad (1.4.1)$$

Because of the boundary condition (1.1.2) we have omitted the terms corresponding to $j = 0$ and $j = n$ from the beginning. It is evident that the function given by (1.4.1) consists of polygonal lines as shown in Fig. 1.2. A function of this shape is called a *piecewise linear polynomial*. The expansion (1.4.1) satisfies

$$\hat{u}_n(x_j) = a_j \qquad (1.4.2)$$

at the node $x = x_j$; that is, the coefficient a_j of the expansion is equal to the value of $\hat{u}_n(x)$ at the node, which is quite convenient for practical purposes.

Now when we try to apply the procedure stated in the previous section starting with the expansion (1.4.1), we encounter an unfavorable situation. That is, although $\hat{u}_n(x)$ can be differentiated once, it cannot be differentiated twice. The first derivative of $\hat{u}_n(x)$ is discontinuous as seen from (1.3.4), and when we try to differentiate further, the *Dirac δ-function* appears at every node in the second derivative of $\phi_j(x)$. It is evident that this is not consistent with (1.1.1).

In order to avoid this inconsistency we consider an equation of the following form instead of (1.1.1):

$$\int_0^1 \left(p \frac{d\hat{u}_n}{dx} \frac{d\hat{\varphi}_j}{dx} + q\hat{u}_n\hat{\varphi}_j \right) dx = \int_0^1 f\hat{\varphi}_j \, dx \qquad j = 1, 2, \ldots, n-1$$

$$(1.4.3)$$

That is, we start with (1.2.6) which is obtained by multiplying (1.1.1) by $\hat{\phi}_j$ followed by integration by parts.

We substitute (1.4.1) for $\hat{u}_n$ on the left-hand side of (1.4.3) and carry out termwise integration using (1.3.3) or (1.3.4). Then, corresponding to each integral we have

Fig. 1.2 Piecewise linear polynomial $\hat{u}_n(x)$.

$$\int_0^1 \hat{\varphi}_k(x)\hat{\varphi}_j(x)\ dx = \begin{cases} 0 & j < k-1 \\[1mm] \dfrac{1}{6}h & j = k-1 \\[1mm] \dfrac{2}{3}h & j = k \\[1mm] \dfrac{1}{6}h & j = k+1 \\[1mm] 0 & j > k+1 \end{cases} \tag{1.4.4}$$

$$\int_0^1 \frac{d\varphi_k}{dx}\frac{d\hat{\varphi}_j}{dx}\ dx = \begin{cases} 0 & j < k-1 \\[1mm] -\dfrac{1}{h} & j = k-1 \\[1mm] \dfrac{2}{h} & j = k \\[1mm] -\dfrac{1}{h} & j = k+1 \\[1mm] 0 & j > k+1 \end{cases} \tag{1.4.5}$$

Substitution of these values into (1.2.6) leads to the following system of linear equations with respect to $\{a_j\}$:

$$(K + M)a = f \tag{1.4.6}$$

where

$$a = \begin{pmatrix} a_1 \\ a_2 \\ \vdots \\ a_{n-1} \end{pmatrix} \tag{1.4.7}$$

and, if we define

$$f_j = \int_0^1 f(x)\hat{\varphi}_j(x)\ dx \tag{1.4.8}$$

then

$$f = \begin{pmatrix} f_1 \\ f_2 \\ \vdots \\ f_{n-1} \end{pmatrix} \tag{1.4.9}$$

K and M are $(n-1) \times (n-1)$ matrices defined as

$$K = \frac{p}{h} \begin{pmatrix} 2 & -1 & & & & \\ -1 & 2 & -1 & & 0 & \\ & -1 & 2 & \ddots & & \\ & & \ddots & \ddots & \ddots & \\ 0 & & & \ddots & 2 & -1 \\ & & & & -1 & 2 \end{pmatrix} \qquad (1.4.10)$$

$$M = \frac{qh}{6} \begin{pmatrix} 4 & 1 & & & & \\ 1 & 4 & 1 & & 0 & \\ & 1 & 4 & \ddots & & \\ & & \ddots & \ddots & \ddots & \\ 0 & & & \ddots & 4 & 1 \\ & & & & 1 & 4 \end{pmatrix} \qquad (1.4.11)$$

In accordance with the traditional terminology of structural mechanics K and M are called the *stiffness matrix* and the *mass matrix,* respectively.

1.5 *Properties of Matrices and the Finite Element Solution*

One of the most remarkable features of the matrices mentioned above is that they are *tridiagonal.* A tridiagonal matrix is a matrix whose entries vanish except on the diagonal and on the subdiagonal. In the first example (1.1.3), in which trigonometric functions were used as basis functions that were orthogonal to each other along with their derivatives, the matrices K and M were both diagonal. On the other hand, in the present example in which the basis functions (1.3.3) are used, the matrices are not completely diagonal but *nearly diagonal,* that is, tridiagonal. It is evident that the reason why they are nearly diagonal is that the support of the basis functions (1.3.3) is localized in a very small domain. Generally speaking, if functions whose support is localized in a small domain are used as basis functions, the nonzero entries of the coefficient matrix of the system of linear equations will be concentrated close to the diagonal, although not completely on the diagonal. As will be seen later, this matrix pattern leads to highly efficient numerical computation and substantial saving of computer memory.

The coefficient matrices K and M of the system of linear equations (1.4.6) are *symmetric.* In addition, K is *positive definite* in the present case, because for any vector

$$b = \begin{pmatrix} b_1 \\ b_2 \\ \vdots \\ b_{n-1} \end{pmatrix} \neq 0 \qquad (1.5.1)$$

we have

$$b^T K b = \int_0^1 p \left(\sum_{j=1}^{n-1} b_j \frac{d\hat{\varphi}_j}{dx} \right)^2 dx > 0 \qquad (1.5.2)$$

since p is assumed to be positive. Here b^T is the transposed vector of b. The matrix M is also shown to be positive definite in the same way. When p or q is a function of x, K and M also become positive definite provided that $p(x) > 0$ and $q(x) > 0$. If K and M are positive definite, then $K + M$ is also positive definite, so that (1.4.6) can be solved for $\{a_j\}$. Then, if we substitute $\{a_j\}$ into (1.4.1), we have an approximate solution $\hat{u}_n(x)$. This solution is just the *finite element solution,* which is the main topic of this book. It will become clear in Chap. 3 in what sense it is an approximation to an exact solution of the problem.

Although the example stated above is a simple model problem in one dimension, the same idea applies directly to problems in two or three dimensions. For example, in a two-dimensional problem, we divide the given domain into small triangles and choose piecewise polynomial basis functions each of which vanishes except in a small number of triangles adjacent to each other. Then we construct an approximate solution in terms of a linear combination of these basis functions and apply Galerkin's method.

The method stated above in which we divide the whole domain into subdomains, for example, triangles with (not an infinitely small but) a finite area, choose basis functions such that each of them does not vanish only in the near neighborhood of a particular node, construct an approximate solution to the given problem in terms of a linear combination of piecewise polynomials, and apply the Galerkin method, is generically called the finite element method (FEM).

In the FEM we usually must solve a large system of linear equations such as (1.4.6). For this reason no one thought of applying the FEM as a powerful tool for solving practical problems until high-speed computers with large storage capacities became available.

2

The Variational Principle and Galerkin's Method

2.1 *Functionals of Variation and Bilinear Forms*

As stated in the previous chapter, the FEM is a type of Galerkin method in which the support of the basis functions is localized. The FEM can also be formulated through the variational principle, and in this section we present an example of how this can be done. The problem is again a boundary value problem with a boundary condition

$$u(0) = u(1) = 0 \qquad (2.1.1)$$

under which a *functional*

$$J[u] = \frac{1}{2} \int_0^1 (pu'^2 + qu^2 - 2fu) \ dx \qquad (2.1.2)$$

is minimized. Here p, q, and f are given functions of x, and p and q are assumed to satisfy the following inequalities:

$$p_M \geq p(x) \geq p_m > 0 \qquad (2.1.3)$$

$$q_M \geq q(x) \geq q_m \geq 0 \qquad (2.1.4)$$

In this chapter we will derive the same result as that obtained in the preceding chapter by minimizing the functional $J[u]$. Before doing so we introduce some notation that will be used when formulating the FEM in a more mathematically rigorous way. First we define a *bilinear form*

$$a(u,v) = \int_0^1 (pu'v' + quv)\, dx \qquad (2.1.5)$$

corresponding to the main term of the derivatives in $J[u]$, that is, to the first two terms of the integral in (2.1.2). Here u' and v' are the derivatives of u and v, respectively. The bilinear form $a(u,v)$ satisfies

$$a(\alpha u + \beta w, v) = \alpha a(u, v) + \beta a(w, v) \qquad (2.1.6)$$

$$a(u, \alpha v + \beta w) = \alpha a(u, v) + \beta a(u, w) \qquad (2.1.7)$$

by definition, where α and β are constants. As defined above, $a(u,v)$ is symmetric; that is,

$$a(u,v) = a(v,u) \qquad (2.1.8)$$

Next we define

$$(f,u) = \int_0^1 fu\, dx \qquad (2.1.9)$$

corresponding to the last term of the integral in (2.1.2). Then (2.1.2) can be written as

$$J[u] = \tfrac{1}{2}a(u,u) - (f,u) \qquad (2.1.10)$$

2.2   The H_1 Norm and Admissible Functions

In order for the minimization of $J[u]$ to make sense the right-hand side of (2.1.2) must be defined properly. That is, u must be differentiable at least once, and its square must be integrable. Therefore we fix the interval at $[0,1]$ and consider the set of functions such that its *norm*

$$\|u\|_1 = \left[\int_0^1 (u^2 + u'^2)\, dx\right]^{1/2} \qquad (2.2.1)$$

is finite.

In general, the norm of a set of functions should satisfy the following three conditions:

1. $\|u\| \geq 0$, where the equality holds if and only if $u \equiv 0$ (2.2.2)
2. For real α, $\|\alpha u\| = |\alpha|\,\|u\|$ (2.2.3)
3. $\|u + v\| \leq \|u\| + \|v\|$ (2.2.4)

It is evident that the norm (2.2.1) satisfies conditions 1 and 2. It will be shown in Sec. 3.2 that it also satisfies condition 3.

The definition of the norm (2.2.1) has u' together with u in its integrand. It should be noted that, if the norm is defined without u', that is, by

$$\|u\| = \left[\int_0^1 u'^2 \, dx \right]^{1/2} \tag{2.2.5}$$

then $\|u\| = 0$ holds when $u' = 0$; that is, $u = $ constant and condition 1 is not satisfied. If, on the other hand, u is restricted to functions that satisfy $u(0) = 0$ or $u(1) = 0$, (2.2.5) can be a norm. A quantity that satisfies conditions 2 and 3 but does not satisfy condition 1 in the sense that $\|u\| = 0$ may hold even when $u \neq 0$, such as (2.2.5), is called a *seminorm*. In Sec. 7.4 an error analysis will be given based on a seminorm.

Since we have defined the norm of a function, we next introduce a *function space*. As a natural characteristic of a space, if two elements u and v belong to a function space, then the sum $u + v$ and its constant multiple cu should also belong to the same function space.

The function space we will mainly consider is a space whose elements are such that their first derivative is square-integrable, for example, the space in which the norm is defined by (2.2.1). This space is denoted by H_1 ($[0,1]$), abbreviated as H_1, and its norm is called the H_1 *norm*. The subscript 1 indicates that the element is differentiable once. In general, also in the case of functions with many variables, the function space in which a norm like (2.2.1) is defined, which includes derivatives of order up to m, is called the *Sobolev space* and is denoted by H_m. The norm is called the *Sobolev norm*. The H_1 norm defined above is one of the simplest examples of the Sobolev space. In the case of functions with one variable the space whose norm is defined by

$$\|u\|_m = \left[\int_0^1 (u^2 + u'^2 + \cdots + u^{(m)2}) \, dx \right]^{1/2} \tag{2.2.6}$$

is H_m.

We denote by $\overset{\circ}{H}_1$ the subspace of H_1 that consists of functions satisfying

$$u(0) = u(1) = 0 \tag{2.2.7}$$

Then, in a variational problem in which the functional (2.1.10) is minimized under the constraint (2.1.1), we can use an element of $\overset{\circ}{H}_1$. In other words, the *admissible function* for our variational problem is an element of $\overset{\circ}{H}_1$.

If the boundary condition is not homogeneous like (2.2.7) but is given, for example, as

$$u(0) = u_0 \neq 0 \tag{2.2.8}$$

the set of all functions in H_1 satisfying this condition is not a subspace of H_1. The reason is that the boundary value of the sum $u + v$ of such functions is $2u_0$, so that the original boundary condition is not satisfied. That is, the subspace must have the same characteristics as the space.

2.3 The First Variation

In order to minimize the functional $J[u]$ in (2.1.10) we write

$$
\begin{aligned}
u_\epsilon &= u + \epsilon\eta \\
&= u + \delta u
\end{aligned}
\tag{2.3.1}
$$

where u = function that minimizes $J[u]$

ϵ = an arbitrary real number

η = an arbitrary function belonging to $\mathring{H}_1$

$$
\delta u = \epsilon\eta(x) \tag{2.3.2}
$$

is called the *variation* of the function u, and u_ϵ is an admissible function. In general a function such as u_ϵ having a parameter through which the functional $J[u]$ is minimized is called a *test function*. Incidentally, it should be noted that the method developed below also applies when the boundary condition is of the inhomogeneous Dirichlet type; that is,

$$
u(0) = u_0 \qquad u(1) = u_1 \tag{2.3.3}
$$

where either u_0 or u_1 does not vanish. Also, in this case η must be an element of $\mathring{H}_1$, because u on the right-hand side of (2.3.1) satisfies (2.3.3) from the beginning.

Now we substitute the test function u_ϵ for u in (2.1.10). If we note that $a(u,v)$ is a symmetric bilinear form, we have

$$
\begin{aligned}
J[u_\epsilon] &= \tfrac{1}{2}a(u + \epsilon\eta, u + \epsilon\eta) - (f, u + \epsilon\eta) \\
&= \tfrac{1}{2}a(u, u) - (f, u) + \epsilon\{a(u, \eta) - (f, \eta)\} + \tfrac{1}{2}\epsilon^2 a(\eta, \eta) \\
&= J[u] + \epsilon\{a(u, \eta) - (f, \eta)\} + \tfrac{1}{2}\epsilon^2 a(\eta, \eta)
\end{aligned}
\tag{2.3.4}
$$

In view of the fact that $J[u_\epsilon]$ is a function of ϵ we write

$$
\Phi(\epsilon) = J[u_\epsilon] \tag{2.3.5}
$$

Then, since u minimizes $J[u]$, $\Phi(\epsilon)$ attains its stationary point when $\epsilon = 0$. That is,

$$
\Phi'(0) = 0 \tag{2.3.6}
$$

Therefore we have

$$
\left.\frac{\partial}{\partial\epsilon} J[u_\epsilon]\right|_{\epsilon=0} = 0 \tag{2.3.7}
$$

From this equation

$$a(u, \eta) - (f, \eta) = 0 \qquad \forall \eta \in \mathring{H}_1 \qquad (2.3.8)$$

holds. In the present example this becomes

$$\int_0^1 (pu'\eta' + qu\eta - f\eta) \, dx = 0 \qquad \forall \eta \in \mathring{H}_1 \qquad (2.3.9)$$

In place of the expression, "for any η belonging to $\mathring{H}_1$," we will hereafter write

$$\forall \eta \in \mathring{H}_1 \qquad (2.3.10)$$

in accordance with a conventional expression in mathematics.

In general the linear term in $J[u_\epsilon]$ with respect to ϵ

$$\delta J[u] \equiv \epsilon \Phi'(0) \qquad (2.3.11)$$

is called the *first variation* of $J[u]$. Also, the term that includes ϵ^2 is called the *second variation*. As stated above, the necessary condition for $J[u]$ to attain its stationary point is that the first variation vanishes there.

Integrating (2.3.9) by parts and taking into account $\eta \in \mathring{H}_1$, that is, $\eta(0) = \eta(1) = 0$, we have

$$\int_0^1 \{-(pu')' + qu - f\}\eta \, dx = 0 \qquad \forall \eta \in \mathring{H}_1 \qquad (2.3.12)$$

2.4 The Fundamental Theorem of Variation and Euler's Equation

Let $g(x)$ be a continuous function in $[a,b]$ and suppose that for any function $\zeta(x)$ satisfying the boundary condition

$$\zeta(a) = \zeta(b) = 0 \qquad (2.4.1)$$

it holds that

$$\int_a^b g(x)\zeta(x) \, dx = 0 \qquad (2.4.2)$$

Then we have

$$g(x) \equiv 0 \qquad (2.4.3)$$

The reason is as follows. If $g(x_0) > 0$ at some x_0 in $[a,b]$, then $g(x) > 0$ in some neighborhood of x_0 because $g(x)$ is continuous. Then, if we choose a function $\zeta(x)$ which is positive in this neighborhood and zero otherwise, the left side of (2.4.2) becomes positive, which is a contradiction. The same situation holds if we assume that $g(x)$ is negative at some point. It is evident that the above discussion can be generalized to variational problems in more than one dimension. The theorem stated above that concludes (2.4.3) from the fact that for any function satisfying the condition (2.4.1) the equation (2.4.2) holds is called the *fundamental theorem of variation*.

Again we return to (2.3.12). If we assume that p, q, and f satisfy appropriate conditions and that $-(pu')' + qu - f$ is continuous as a whole, we can conclude from the fundamental theorem of variation that

$$-(pu')' + qu - f = 0 \tag{2.4.4}$$

This is called the *Euler's equation* corresponding to the functional (2.1.2).

Comparing the Euler's equation and (1.1.1) we see that they are equal to each other, except that p and q are constants in (1.1.1). Thus we conclude that, if p, q, and f satisfy some appropriate continuous conditions, (1.1.1) can be obtained by making $J[u]$ in (2.1.2) stationary.

2.5 Positive Definite Bilinear Forms

The integral (2.3.9) is an equation with respect to the unknown function u. This is the basic equation of the FEM, the details of which will be given later. If the functional $J[u]$ is required by u to attain not only its stationary state but also its minimum it is necessary for the coefficient $a(\eta,\eta)$ of ϵ^2 in the second variation in (2.3.4) to be positive. In order to describe this situation in a function analytic fashion we introduce the notion of positive definiteness of the bilinear form $a(u,v)$. To begin with we present an important inequality.

For later convenience we generalize the interval of integration $(0,1)$ to (α,β). Suppose that the function u belongs to H_1 defined over (α,β) and satisfies

$$u(\alpha) = u(\beta) = 0 \tag{2.5.1}$$

that is, $u \in \mathring{H}_1$.

Since u vanishes at both end points, it can be expanded in terms of a Fourier sine series:

$$u(x) = \sum_{n=1}^{\infty} b_n \sin \frac{n\pi(x-\alpha)}{l} \qquad l = \beta - \alpha \tag{2.5.2}$$

Then it is easy to show that

$$\int_{\alpha}^{\beta} u^2 \, dx = \frac{l}{2} \sum_{n=1}^{\infty} b_n^2 \tag{2.5.3}$$

and

$$\int_\alpha^\beta u'^2\, dx = \frac{l}{2} \sum_{n=1}^\infty \left(\frac{n\pi}{l}\right)^2 b_n^2 \tag{2.5.4}$$

On the other hand, by virtue of

$$b_n^2 \le \frac{l^2}{\pi^2} \left(\frac{n\pi}{l}\right)^2 b_n^2 \tag{2.5.5}$$

for $n \ge 1$ we have the following inequality from (2.5.3) and (2.5.4):

$$\int_\alpha^\beta u^2\, dx \le \frac{l^2}{\pi^2} \int_\alpha^\beta u'^2\, dx \tag{2.5.6}$$

The equality sign holds when $u(x) = \sin \pi(x - \alpha)/l$.
We assume here for later convenience that

$$\int_\alpha^\beta u''^2\, dx < +\infty \tag{2.5.7}$$

Then, since from (2.5.2) we have

$$\int_\alpha^\beta u''^2\, dx = \frac{l}{2} \sum_{n=1}^\infty \left(\frac{n\pi}{l}\right)^4 b_n^2 \tag{2.5.8}$$

we obtain, as above,

$$\int_\alpha^\beta u'^2\, dx \le \frac{l^2}{\pi^2} \int_\alpha^\beta u''^2\, dx \tag{2.5.9}$$

From (2.5.9) and (2.5.6)

$$\int_\alpha^\beta u^2\, dx \le \frac{l^4}{\pi^4} \int_\alpha^\beta u''^2\, dx \tag{2.5.10}$$

follows. The equality sign also holds when $u(x) = \sin \pi(x - \alpha)/l$. This inequality will be used in Sec. 3.8.

Now we return to the bilinear from $a(u,v)$ in (2.1.5). If we put $v = u$ in (2.1.5), we have from (2.1.3) and (2.1.4)

$$a(u,u) \ge p_m \int_0^1 u'^2\, dx + q_m \int_0^1 u^2\, dx$$

$$\ge p_m \int_0^1 u'^2\, dx = \frac{p_m}{1 + \pi^{-2}} \int_0^1 (1 + \pi^{-2}) u'^2\, dx \tag{2.5.11}$$

From (2.5.6), in which $\alpha = 0$, $\beta = 1$, and $l = 1$, we have

$$a(u,u) \ge \frac{p_m}{1 + \pi^{-2}} \int_0^1 (u'^2 + u^2)\, dx$$

$$= \frac{p_m}{1 + \pi^{-2}} \|u\|_1^2 \qquad \forall u \in \overset{\circ}{H}_1 \tag{2.5.12}$$

We assume here that u belongs to $\mathring{H}_1$ because (2.5.6) is derived under the assumption that (2.5.1) holds.

If, for a bilinear form $a(u,v)$ defined for elements in a function space H, there exists a positive constant γ such that

$$a(v, v) \geq \gamma\|v\|^2 \qquad \forall v \in H \qquad (2.5.13)$$

$a(u,v)$ is said to be *positive definite* or *coersive* or *elliptic*. We use the term *H-elliptic* to indicate the function space explicitly, and $\|v\|$ is the norm in H. We consider this condition elliptic because the term of the highest-order derivative of an elliptic boundary value problem in more than one space dimension satisfies this condition.

As seen from (2.5.12), the bilinear form defined by (2.1.5) is H_1-elliptic. Therefore $a(\eta, \eta) > 0$ holds for any nonzero function η belonging to $\mathring{H}_1$, so that the second variation in (2.3.4) is positive, hence $J[u_\epsilon]$ attains its minimum at $\epsilon = 0$.

The assumption (2.1.4) includes the case $q(x) \equiv 0$. We see in this case also that $a(\eta, \eta) > 0$ for any $\eta \neq 0$. Note, however, that the inequality holds because η belongs to $\mathring{H}_1$. When, on the other hand, η belongs to H_1, $a(\eta, \eta)$ may vanish if we take $\eta = \text{constant} \neq 0$.

2.6   Weak Forms

As we have already seen, when p, q, and f satisfy an appropriate continuity condition, the problem in which the functional (2.1.10) is minimized under the constraint condition (2.1.1) is equivalent to the problem given by (1.1.1) and (1.1.2). However, if the continuity condition is not assumed, the two problems are not always equivalent to each other. Equation (2.3.9), however, is equivalent to the minimizing problem.

Equation (2.3.9) is called a *weak form* or a *variational equation* corresponding to the problem (1.1.1) and (1.1.2). As will be seen in Sec. 2.8, the weak form is employed as the basic equation in the FEM. The weak form is sometimes called the *principle of virtual work*.

While (2.4.4) with (2.1.1) implies (2.3.9), (2.3.9) does not, as seen above, imply (2.4.4). Therefore (2.3.9) describes a wider problem than (2.4.4). That is, there exists a solution of (2.3.9) that does not satisfy (2.4.4) in the sense that it is not twice-differentiable. In this sense the solution of (2.3.9) is called a *weak solution*.

2.7   A Problem With a Point Load

In this section we consider as an example problem (2.4.4) in which f is a point load at $x = \frac{1}{2}$. Here we let $p = 1$ and $q = 4$. According to a conventional notion of physics we express the point load in terms of the Dirac δ-

function. Then the equations of the present problem and the boundary condition are

$$-\frac{d^2 u}{dx^2} + 4u = -\delta\left(x - \frac{1}{2}\right) \tag{2.7.1}$$

$$u(0) = u(1) = 0 \tag{2.7.2}$$

The Dirac δ-function $\delta(x - \xi) = \delta_\xi$ that corresponds to the point load at $x = \xi$ is defined as a *continuous linear functional*

$$\int \delta(x - \xi)\varphi(x)\ dx = \varphi(\xi) \tag{2.7.3}$$

for any sufficiently smooth function $\phi(x)$ that vanishes except in the neighborhood of $x = \xi$. That is, the δ-function is a continuous linear functional that gives the value of each function at ξ in the set of all sufficiently smooth functions with local support. Taking this into account we multiply a function $\phi(x)$ satisfying $\phi(0) = \phi(1)$ and having the first derivative on both sides of (2.7.1) and integrate it. Then we see that the problem (2.7.1) and (2.7.2) is eventually reduced to a problem in which for any such $\phi(x)$ a solution u satisfying

$$\int_0^1 \left(\frac{du}{dx}\frac{d\varphi}{dx} + 4u\varphi\right) dx = -\varphi\left(\frac{1}{2}\right) \tag{2.7.4}$$

$$u(0) = u(1) = 0 \tag{2.7.5}$$

is sought. It is easy to see that the solution of (2.7.4) and (2.7.5) is given as follows (Fig. 2.1)

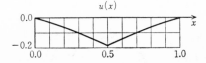

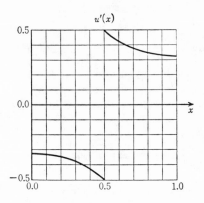

Fig. 2.1 The solution u and its derivative u' for a problem with a point load.

$$u(x) = \begin{cases} -\dfrac{1}{4\cosh 1}\sinh 2x & 0 \le x \le \dfrac{1}{2} \\[2ex] -\dfrac{1}{4\cosh 1}\sinh 2(1-x) & \dfrac{1}{2} < x \le 1 \end{cases} \qquad (2.7.6)$$

Conversely, let us start from the solution $u(x)$. If we substitute it on the left-hand side of (2.7.1), the first derivative becomes discontinuous at $x = \frac{1}{2}$ (Fig. 2.1), so that $u(x)$ does not have a second derivative at $x = \frac{1}{2}$ in the sense of the elementary definition. In the second-order differential equation the order of differentiability of the function $f(x)$ on the right-hand side may be lower by 2 than that of the solution $u(x)$. Thus, also in the present case where the discontinuous function is differentiated, a δ-function appears on the right-hand side so as to make the equation consistent with respect to the order of derivatives. In the present problem the solution (2.7.6) of (2.7.4) and (2.7.5) corresponds to the weak solution of (2.7.1) and (2.7.2).

2.8 Galerkin's Method

Now we return to the FEM. If we try to solve (2.7.1) numerically, we will find it very difficult because (2.7.1) includes the second-order derivative of u and the δ-function. On the other hand, the weak form (2.7.4) has only the first-order derivative, so that it is much easier to solve numerically. Thus, in the FEM a weak form such as (2.7.4) is generally employed as a basic equation.

First we express an approximate solution $\hat{u}_n(x)$ in terms of a linear combination of some appropriate basis functions, for example, of $\{\hat{\phi}_j\}$ defined by (1.3.3):

$$\hat{u}_n(x) = \sum_{j=1}^{n-1} a_j\hat{\varphi}_j(x) \qquad (2.8.1)$$

Note that (2.8.1) satisfies the boundary condition

$$\hat{u}_n(0) = \hat{u}_n(1) = 0 \qquad (2.8.2)$$

because the terms corresponding to the nodes $j = 0$ and $j = n$ are not included in (2.8.1). Next we choose $\hat{\phi}_j$, $j = 1,2,\ldots,n-1$, in (1.3.3) for $\phi(x)$ in (2.7.4). Substituting $\hat{u}_n$ for u and $\hat{\phi}_j$ for ϕ in (2.7.4), we immediately obtain

$$\int_0^1 \left(\frac{d\hat{u}_n}{dx}\frac{d\hat{\varphi}_j}{dx} + 4\hat{u}_n\hat{\varphi}_j\right) dx = -\hat{\varphi}_j\left(\frac{1}{2}\right) \qquad j = 1,2,\ldots,n-1$$

$$(2.8.3)$$

This is just (1.4.3), and the problem eventually becomes a problem of solving the system of linear equations (1.4.6), that is,

$$\sum_{i=1}^{n-1} (K_{ji} + M_{ji}) a_i = -\hat{\varphi}_j\left(\frac{1}{2}\right) \qquad j = 1, 2, \ldots, n-1 \qquad (2.8.4)$$

where K_{ij} and M_{ij} are the ij entries of the matrices (1.4.10) and (1.4.11), respectively, and $\hat{\varphi}_j(\frac{1}{2})$ is 1 when $j = n/2$ and 0 otherwise, provided that n is even. Specifically, we take $n = 10$ and solve the problem numerically. Then we will obtain an approximate solution that coincides with the exact solution (2.7.6) in up to four significant digits at every node.

We summarize the procedure given above in a more generalized fashion so that we can apply it to other problems. Let $a(u,v)$ be a given bilinear form like (2.1.5) and let the given problem be

$$a(u, v) - (f, v) = 0 \qquad \forall v \in \mathring{H}_1$$
$$u(0) = u(1) = 0 \qquad\qquad (2.8.5)$$

We fix the nodes of (1.3.3) and denote the space of all piecewise linear functions (2.8.1) expressed in terms of a linear combination of (1.3.3) by $\mathring{K}_n$. Note that each of the $n - 1$ basis functions $\hat{\phi}_1, \hat{\phi}_2, \ldots, \hat{\phi}_{n-1}$ is fixed since we have fixed the nodes. Therefore $\mathring{K}_n$ becomes a finite $(n-1)$-dimensional subspace of $\mathring{H}_1$. Then we replace u and v with $\hat{u}_n$ and $\hat{v} \in \mathring{K}_n$, respectively, in the weak form (2.8.5), so that we have

$$a(\hat{u}_n, \hat{v}) - (f, \hat{v}) = 0 \qquad \forall \hat{v} \in \mathring{K}_n \qquad (2.8.6)$$
$$\hat{u}_n(0) = \hat{u}_n(1) = 0 \qquad\qquad (2.8.7)$$

Note that any $\hat{v} \in \mathring{K}_n$ can be expressed in terms of a linear combination of $\hat{\phi}_j$, $j = 1, 2, \ldots, n-1$. Then the solution of (2.8.6) and (2.8.7) is reduced to the solution of

$$a(\hat{u}_n, \hat{\varphi}_j) - (f, \hat{\varphi}_j) = 0 \qquad j = 1, 2, \ldots, n-1 \qquad (2.8.8)$$
$$\hat{u}_n(0) = \hat{u}_n(1) = 0 \qquad\qquad (2.8.9)$$

that is, the system of linear equations (1.4.6).

As already mentioned, Galerkin's method is a method in which the weak form (2.8.5) is approximated as (2.8.6) and (2.8.7) and the approximate equation then solved numerically. A type of Galerkin's method in which piecewise polynomials are employed as the basis functions $\hat{\phi}_j$ is the FEM, and (2.8.6) is called the *Galerkin equation*.

Equations encountered in practical problems often have singularities, such as the point singularity of f in (2.7.1), or some kind of discontinuity in $p(x)$ or $q(x)$. In such cases Galerkin's method gives us a reasonable approximate solution like (2.7.6) in a natural way. Also, it is quite convenient from the standpoint of numerical computation that for basis functions only first-order, and not second-order, differentiability is required.

2.9   *The Ritz Method*

Suppose in general that the functional $J[u]$ has a minimum as a function of u. Then we can obtain an approximate solution by first constructing a test function with unknown parameters, then substituting it for u in $J[u]$, and finally determining the parameters by directly minimizing $J[u]$. This is called the *direct method* or the *Ritz method*.

We apply the direct method to problem (2.1.2). As an approximate solution we employ the piecewise polynomial

$$\hat{u}_n(x) = \sum_{j=1}^{n-1} a_j \hat{\varphi}_j(x) \qquad (2.9.1)$$

given in the preceding section. The coefficients $\{a_j\}$ are the parameters to be determined. Note that $\hat{u}_n(x)$ as given above satisfies the boundary condition (2.1.1). Substituting $\hat{u}_n(x)$ for u in the functional (2.1.2) to be minimized, we have a *quadratic form* with respect to $\{a_j\}$:

$$J[\hat{u}_n] = \frac{1}{2} \int_0^1 \left\{ p\left(\sum_{j=1}^{n-1} a_j \frac{\partial \hat{\varphi}_j}{\partial x} \right)^2 + q\left(\sum_{j=1}^{n-1} a_j \hat{\varphi}_j \right)^2 - 2f \sum_{j=1}^{n-1} a_j \hat{\varphi}_j \right\} dx$$

$$= \frac{1}{2} a^T (K + M) a - f^T a \qquad (2.9.2)$$

The vectors a and f and the matrices K and M are defined in Sec. 1.4.

Since as already seen in (1.5.2) the matrix $K + M$ is positive definite, the problem of finding a stationary value of $J[\hat{u}_n]$ is equivalent to the problem of finding a minimum of $J[\hat{u}_n]$. This is of course due to the positive definiteness (2.5.12) of the bilinear form $a(u,v)$ in (2.1.5). The condition that $J[\hat{u}_n]$ be a minimum can be written as

$$\frac{\partial J[\hat{u}_n]}{\partial a_i} = \sum_{j=1}^{n-1} (K_{ij} + M_{ij}) a_j - f_i = 0 \qquad (2.9.3)$$

where K_{ij} and M_{ij} are the ij entries of K and M, respectively. As expected, this equation is equivalent to (1.4.6) obtained by Galerkin's method. Thus we can conclude that, in general, if the bilinear form $a(u,v)$ being considered is positive definite, then the result obtained with the Ritz method based on the variational principle and the result obtained with Galerkin's method are equivalent.

2.10   *The Inhomogeneous Dirichlet Boundary Condition*

We can treat the inhomogeneous Dirichlet boundary condition

$$u(0) = u_0 \qquad u(1) = u_1 \qquad (2.10.1)$$

in which either u_0 or u_1 does not vanish, in almost the same way as the homogeneous case. In the inhomogeneous case we only have to change the boundary condition and solve

$$a(u,v) - (f,v) = 0 \qquad \forall v \in \mathring{H}_1 \tag{2.10.2}$$

$$u(0) = u_0 \qquad u(1) = u_1 \tag{2.10.3}$$

As already seen in Sec. 2.3, where we discussed the first variation, v in (2.10.2) corresponds essentially to the variation δu of u, hence we can take any element belonging to $\mathring{H}_1$ as v.

2.11 The Natural Boundary Condition

The boundary conditions we have treated so far are of the Dirichlet type. In this section we use another type of boundary condition and consider the following problem:

$$-\frac{d}{dx}\left(p\,\frac{du}{dx}\right) + qu = f \qquad 0 < x < 1 \tag{2.11.1}$$

$$\alpha u'(0) - \beta u(0) = 0 \tag{2.11.2}$$

$$\gamma u'(1) + \delta u(1) = 0 \tag{2.11.3}$$

We assume that α, β, γ, and δ are constants such that $\alpha\beta \geq 0$, $\gamma\delta \geq 0$ and $\alpha \neq 0$, $\gamma \neq 0$. Also, we assume that p and q are functions of x satisfying (2.1.3) and (2.1.4).

Multiplying both sides of (2.11.1) by an arbitrary function $v \in H_1$ and carrying out integration by parts over $(0,1)$ using (2.11.2) and (2.11.3), we have

$$\int_0^1 (pu'v' + quv)\;dx + \frac{\beta}{\alpha}p(0)u(0)v(0) + \frac{\delta}{\gamma}p(1)u(1)v(1)$$

$$= \int_0^1 fv\;dx \qquad \forall v \in H_1 \tag{2.11.4}$$

Note that $v(0) = v(1) = 0$ is not assumed here. Corresponding to this problem we define a bilinear form with respect to u, $v \in H_1$:

$$a(u, v) = \int_0^1 (pu'v' + quv)\;dx + \frac{\beta}{\alpha}p(0)u(0)v(0) + \frac{\delta}{\gamma}p(1)u(1)v(1) \tag{2.11.5}$$

Then we can write (2.11.4) as

$$a(u,v) - (f,v) = 0 \qquad \forall v \in H_1 \tag{2.11.6}$$

This is the weak form corresponding to the boundary value problem (2.11.1) through (2.11.3).

In contrast with the former boundary condition $u(0) = u(1) = 0$, we need not take into account the boundary condition (2.11.2) and (2.11.3) when solving the weak form (2.11.6). The reason is as follows. If we integrate (2.11.4) by parts toward the original form we have

$$\int_0^1 (-(pu')' + qu - f)v \, dx$$

$$-\frac{1}{\alpha} p(0)\{\alpha u'(0) - \beta u(0)\}v(0) + \frac{1}{\gamma} p(1)\{\gamma u'(1) + \delta u(1)\}v(1) = 0$$

$$(2.11.7)$$

First, for $v \in H_1$ in (2.11.7) we choose an arbitrary function v satisfying $v(0) = v(1) = 0$, which gives (2.11.1) from the fundamental theorem of variation applied to the integral in (2.11.7). Thus the integral in (2.11.7) vanishes. Next, we choose an arbitrary function v satisfying $v(0) \neq 0$ and $v(1) = 0$, which gives (2.11.2). Finally, we choose an arbitrary function v satisfying $v(0) = 0$ and $v(1) \neq 0$, which gives (2.11.3).

The situation stated above is well known in the field of variation. The functional to be minimized corresponding to the boundary value problem (2.11.1) through (2.11.3) is given by

$$J[u] = \frac{1}{2} \int_0^1 (pu'^2 + qu^2 - 2fu) \, dx + \frac{\beta}{2\alpha} p(0) u^2(0) + \frac{\delta}{2\gamma} p(1) u^2(1)$$

$$= \frac{1}{2} a(u, u) - (f, u) \qquad (2.11.8)$$

In order to derive it, as before, we let

$$u_\epsilon = u + \epsilon\eta \qquad (2.11.9)$$

where ϵ is an arbitrary real number and η is an arbitrary function belonging to H_1. Substituting this expression for u in (2.11.8) and letting

$$\frac{\partial}{\partial \epsilon} J[u_\epsilon]\bigg|_{\epsilon=0} = 0 \qquad (2.11.10)$$

we have

$$a(u, \eta) - (f, \eta) = 0 \qquad \forall \eta \in H_1 \qquad (2.11.11)$$

which is the weak form corresponding to (2.3.8). From this we obtain the boundary conditions (2.11.2) and (2.11.3) as shown above.

Thus, if the boundary condition is given in the form of (2.11.2) and (2.11.3), we need not impose it on the trial function when solving the corresponding weak form. Since the condition (2.11.2) and (2.11.3) is naturally satisfied by the solution of the weak form, it is called the *natural boundary condition*. The fact that we need not take into account the boundary condition is quite convenient from the standpoint of numerical computation. The boundary condition

$$u'(0) = u'(1) = 0 \qquad\qquad (2.11.12)$$

is called the *Neumann condition* and is a typical example of a natural boundary condition.

In contrast with the natural condition, one that must be imposed on a trial function, like the Dirichlet condition, is called the *essential boundary condition*.

In Galerkin's method a function that allows the weak form of a given problem to be computed properly and satisfies the essential boundary condition if it is imposed is also called an *admissible function* after the manner of the variation. When the boundary condition is a natural one, it is not necessary for the admissible function to satisfy it beforehand. It is only necessary that it be a function that allows the given weak form to be computed properly, that is, a function that can be differentiated properly up to the order of derivatives appearing in the integral.

3

Error Analysis of the Finite Element Method in One Dimension

3.1 *Hilbert Space and the Schwarz Inequality*

So far we have treated the FEM as a method for obtaining an approximate solution of a given problem. In this chapter we will raise the question, In what sense is this solution an approximation to the exact solution? That is, the main subject of the present chapter is an error analysis of the finite element solution of a given boundary value problem in one space dimension. Before proceeding to the main subject we present a short introduction to the mathematical notation required for the analysis.

Suppose that an *inner product* (u,v) is defined for any two elements u and v belonging to the function space under consideration. An inner product (u,v) must in general satisfy the following three conditions:

1. $(u,u) \geq 0$, where the equality sign holds if and only if $u = 0$ (3.1.1)
2. $(u,v) = (v,u)$ (3.1.2)
3. For real α and β, $(\alpha u + \beta v) = \alpha(u,w) + \beta(v,w)$ (3.1.3)

If an inner product is defined, the function space becomes a *Hilbert space*. The norm of the Hilbert space is defined by the inner product as

$$\|u\| = \sqrt{(u,u)} \qquad (3.1.4)$$

Although completeness is usually required in the rigorous definition of a Hilbert space, we do not assume explicitly the completeness of the Hilbert space here. When we discriminate an inner product space without completeness from a Hilbert space in the rigorous sense, we call the former a *pre-Hilbert space*.

The space H_1 that we introduced in Sec. 2.2 becomes a Hilbert space if we define the inner product by

$$(u, v)_1 = \int_0^1 (u'v' + uv)\, dx \qquad (3.1.5)$$

If two elements u and v in a Hilbert space H satisfy

$$(u, v) = 0 \qquad (3.1.6)$$

then u and v are said to be *orthogonal*, which we represent as

$$u \perp v \qquad (3.1.7)$$

Suppose that in the Hilbert space H a function ψ such that $\|\psi\| = 1$ and an arbitrary function u are given. Then consider a function v given by

$$v = c\psi \qquad u - v \perp \psi \qquad (3.1.8)$$

(Fig. 3.1) where v is sometimes called an *orthogonal projection* of u in the direction of ψ. The orthogonal relation (3.1.8) can be expressed as

$$(u - c\psi, \psi) = 0 \qquad (3.1.9)$$

If we solve (3.1.9) for c and substitute it into (3.1.8), we have

$$v = (u, \psi)\psi \qquad (3.1.10)$$

Therefore from (3.1.9) we obtain

$$\|u\|^2 = \|u - (u, \psi)\psi + (u, \psi)\psi\|^2$$
$$= \|u - (u, \psi)\psi\|^2 + \|(u, \psi)\psi\|^2 \qquad (3.1.11)$$
$$\geq |(u, \psi)|^2$$

We can choose $v/\|v\|$ for ψ, so that if we substitute it into (3.1.11) we have the following *Schwarz inequality:*

$$|(u, v)| \leq \|u\|\|v\| \qquad (3.1.12)$$

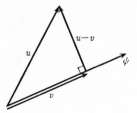

Fig. 3.1 Orthogonal projection v of u in the direction ψ.

The equality sign holds if and only if $u = (u,v)v$, that is, if u is a constant multiple of v.

For example, if the inner product is defined by

$$(u, v) = \int_a^b u(x)v(x) \, dx \qquad (3.1.13)$$

then the Schwarz inequality becomes

$$\left| \int_a^b u(x)v(x) \, dx \right|^2 \leq \left[\int_a^b \{u(x)\}^2 \, dx \right] \left[\int_a^b \{v(x)\}^2 \, dx \right] \quad (3.1.14)$$

3.2 Boundedness of a Bilinear Form

In relation to the Schwarz inequality we give here a definition of the boundedness of $a(u,v)$. From (2.1.3) and (2.1.4) we see that the bilinear form (2.1.5) satisfies

$$|a(u,v)| \leq M \int_0^1 (|u'v'| + |uv|) \, dx \qquad (3.2.1)$$

where $M = \max (p_M, q_M)$. From the inequality

$$|u'v'| + |uv| \leq (u'^2 + u^2)^{1/2}(v'^2 + v^2)^{1/2} \qquad (3.2.2)$$

and (3.1.14) we have

$$|a(u,v)| \leq M\|u\|_1\|v\|_1 \qquad (3.2.3)$$

If $a(u,v)$ satisfies the inequality (3.2.3) for any u and v in H_1, the bilinear form $a(u,v)$ is said to be *bounded* in H_1. All the bilinear forms appearing in this book are bounded.

When we put $p = q = 1$ in the bilinear form (2.1.5), we see that from (3.2.3) the inner product (3.1.5) satisfies

$$|(u, v)_1| \leq \|u\|_1\|v\|_1 \qquad (3.2.4)$$

This is just the Schwarz inequality (3.1.12). From (3.2.4) we can prove that the norm $\|u\|_1$ satisfies condition 3 in Sec. 2.2 required for a norm.

3.3 Energy Space

Suppose the bilinear form $a(u,v)$ is symmetric and also satisfies the ellipticity condition (2.5.13). Then it is easy to see that

$$\|u\|_a \equiv \sqrt{a(u,u)} \qquad (3.3.1)$$

and

$$(u,v)_a \equiv a(u,v) \tag{3.3.2}$$

qualify as the norm and the inner product, respectively. In particular it can be shown that $\|u\|_a$ satisfies condition 3 in Sec. 2.2 from the Schwarz inequality (3.1.12), that is, from

$$|(u,v)_a| \leq \|u\|_a \|v\|_a \tag{3.3.3}$$

Therefore, by introducing the inner product (3.3.2), a new Hilbert space H_a is defined. The subscript a is attached in order to show explicitly that the space is based on the bilinear form $a(u,v)$.

Since $a(u,v) = \|u\|_a^2$ often corresponds to the energy of the physical system under consideration, H_a is called the *energy space*. Also, (3.3.1) and (3.3.2) are called the *energy norm* and the *energy inner product*, respectively.

3.4 Best Approximation in the Energy Norm

Using the mathematical notation given above we now proceed to an error analysis of the finite element solution of the boundary value problem

$$a(u,v) - (f,v) = 0 \qquad \forall v \in \mathring{H} \tag{3.4.1}$$

When the boundary condition is of the Dirichlet type, it must be imposed on u at the boundary. We denote here the space of all functions for which $a(u,v)$ is properly defined by H. Then $\mathring{H}$ in (3.4.1) is a subspace of H consisting of the elements that vanish at the points where the Dirichlet condition is prescribed. We assume that the bilinear form $a(u,v)$ is bounded:

$$a(u,v) \leq M\|u\|\|v\| \qquad \forall u,v \in H \tag{3.4.2}$$

and also $\mathring{H}$-elliptic:

$$a(v,v) \geq \gamma\|v\|^2 \qquad \forall v \in \mathring{H} \tag{3.4.3}$$

where $\|v\|$ is the norm in H and γ is a positive constant. The assumption (3.4.2) implies that if v belongs to H then it belongs to H_a; that is, $\mathring{H}$ is also a subspace of H_a.

As already shown, (3.4.1) is obtained by minimizing

$$J[u] = \tfrac{1}{2}a(u,u) - (f,u) \tag{3.4.4}$$

under the condition that u satisfies the given Dirichlet boundary condition.

Now in order to search for the finite element solution of (3.4.1) we construct a trial function $\hat{v}_n$ in terms of a linear combination of some appropriate basis functions $\hat{\phi}_j, j = 1, 2, \ldots, n$. To solve (3.4.1) by the FEM using this trial function is equivalent to minimizing

$$J[\hat{v}_n] = \tfrac{1}{2} a(\hat{v}_n, \hat{v}_n) - (f, \hat{v}_n)$$
$$= \tfrac{1}{2}(\hat{v}_n, \hat{v}_n)_a - (f, \hat{v}_n) \tag{3.4.5}$$

under the condition that $\hat{v}_n$ satisfies the given Dirichlet boundary condition. Therefore we see that the following relation holds between the exact solution u of (3.4.1) and the trial function $\hat{v}_n$:

$$\|\hat{v}_n - u\|_a^2 = (\hat{v}_n - u, \hat{v}_n - u)_a$$
$$= (\hat{v}_n, \hat{v}_n)_a - 2(u, \hat{v}_n)_a + (u, u)_a$$
$$= (\hat{v}_n, \hat{v}_n)_a - 2(f, \hat{v}_n) + (u, u)_a \tag{3.4.6}$$
$$= 2J[\hat{v}_n] + (u, u)_a$$

On the other hand, since the finite element solution $\hat{u}_n$ is obtained by minimizing $J[\hat{v}_n]$ under the condition that $\hat{v}_n$ satisfies the Dirichlet condition, we have

$$J[\hat{u}_n] \leq J[\hat{v}_n] \tag{3.4.7}$$

From this inequality and from (3.4.6) we obtain

$$\|\hat{u}_n - u\|_a \leq \|\hat{v}_n - u\|_a \tag{3.4.8}$$

This means that obtaining the approximate solution $\hat{u}_n$ by the FEM is equivalent to minimizing the norm $\|\hat{v}_n - u\|_a$ in the energy space H_a under the condition that $\hat{v}_n$ is expressed in terms of a linear combination of $\{\hat{\phi}_j\}$ and that it satisfies the given Dirichlet boundary condition. In other words, the finite element solution $\hat{u}_n$ can be regarded as the *best approximation* to the exact solution u in the energy norm under the condition stated above.

3.5 The Finite Element Solution and Orthogonal Projection

Consider in general a subspace K of the Hilbert space H. If, for an element u of H, a $w \in K$ satisfies

$$(u - w, v) = 0 \qquad \forall v \in K \tag{3.5.1}$$

then we write

$$w = Pu \tag{3.5.2}$$

and we call w an *orthogonal projection* of u onto the subspace K. Here P

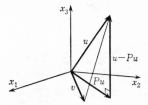

Fig. 3.2 The solution u and its orthogonal projection Pu.

is the *projection operator*. The relation (3.5.1) represents the fact that the residual $u - Pu$ of the orthogonal projection of u onto K is orthogonal to any $v \in K$; that is,

$$u - Pu \perp v \qquad \forall v \in K \tag{3.5.3}$$

Then the residual $u - Pu$ is said to belong to the *orthogonal complement* $K^\perp$ of K.

As seen from Fig. 3.2, if we write the orthogonal projection of a vector $\boldsymbol{u}$ in the three-dimensional $x_1 x_2 x_3$ space onto the two-dimensional $x_1 x_2$ plane as Pu, then $u - Pu$ is orthogonal to any vector $\mathbf{v}$ in the $x_1 x_2$ plane. The situation stated above is a generalization of this orthogonality to the orthogonal relation in the Hilbert space and its subspace.

We apply the idea of orthogonal projection to example (2.1.5). The inner product in the energy space H_a in this case is defined by

$$(u, v)_a = a(u, v) = \int_0^1 (pu'v' + quv) \, dx \tag{3.5.4}$$

under conditions (2.1.3) and (2.1.4), and we use this expression. From the original equation

$$(f, \hat{v}) - a(u, \hat{v}) = 0 \qquad \forall \hat{v} \in \mathring{K}_n \tag{3.5.5}$$

and the approximate equation (2.8.6)

$$(f, \hat{v}) - a(\hat{u}_n, \hat{v}) = 0 \qquad \forall \hat{v} \in \mathring{K}_n \tag{3.5.6}$$

we have

$$a(u, \hat{v}) - a(\hat{u}_n, \hat{v})$$
$$= (u - \hat{u}_n, \hat{v})_a = 0 \qquad \forall \hat{v} \in \mathring{K}_n \tag{3.5.7}$$

If we compare this with (3.5.1), we see that

$$\hat{u}_n = Pu \tag{3.5.8}$$

Since $\mathring{K}_n$ is a subspace of H_a, this result indicates that the finite element solution $\hat{u}_n$ is just the orthogonal projection of the exact solution u in the energy space H_a onto $\mathring{K}_n$.

3.6 *Optimality of the Orthogonal Projection*

When the given Dirichlet boundary condition is homogeneous, the inequality (3.4.8) representing the optimality of the finite element solution
can also be obtained from the orthogonal projection introduced in the preceding section. In this case $\hat{u}_n$ is the orthogonal projection of the exact
solution u in the energy space H_a onto the subspace $\mathring{K}_n$, as seen above.
Therefore, since $\hat{u}_n - \hat{v}_n \in \mathring{K}_n$ for any element $\hat{v}_n$ in $\mathring{K}_n$, we have

$$u - \hat{u}_n \perp \hat{u}_n - \hat{v}_n \tag{3.6.1}$$

That is,

$$(u - \hat{u}_n, \hat{u}_n - \hat{v}_n)_a = 0 \tag{3.6.2}$$

Hence we obtain

$$\|u - \hat{v}_n\|_a^2 = \|u - \hat{u}_n + \hat{u}_n - \hat{v}_n\|_a^2$$
$$= \|u - \hat{u}_n\|_a^2 + \|\hat{u}_n - \hat{v}_n\|_a^2 \geq \|u - \hat{u}_n\|_a^2 \tag{3.6.3}$$

which is just the inequality (3.4.8).

Consider again a vector **u** in the three-dimensional $x_1 x_2 x_3$ space and
its orthogonal projection $Pu = \hat{u}_n$ onto the two-dimensional $x_1 x_2$ plane
shown in Fig. 3.2. Then, by comparing Pu and any vector $v = \hat{v}_n$ in the
$x_1 x_2$ plane, we see that $u - v$ attains its minimum length with $u - Pu$ perpendicular to the $x_1 x_2$ plane. The result shown above may be understood
intuitively from this observation.

When the given Dirichlet boundary condition is inhomogeneous, the
set of all functions in H_a satisfying this condition is not a subspace of
H_a, as mentioned in the last paragraph of Sec. 2.2. However, since the
difference $\hat{u}_n - \hat{v}_n$ in (3.6.2) belongs to $\mathring{K}_n$, (3.5.7) holds for $\hat{v} = \hat{u}_n - \hat{v}_n$
in this case also, so that the result (3.6.3) is true.

3.7 *Interpolation by Piecewise Linear Polynomial*

Although the inequality (3.4.8) can be regarded as an estimate of the
error $\|\hat{u}_n - u\|_a$ of the approximate solution $\hat{u}_n$, it does not give any infor-
mation in practice because it has an arbitrary function $\hat{v}_n$ on the right-
hand side. Suppose, however, that we can choose an appropriate $\hat{v}_n$ very
close to the exact solution u and also that we can obtain an estimate of
its error $\|\hat{v}_n - u\|_a$. Then by (3.4.8) this gives a specific upper bound of
the error $\|\hat{u}_n - u\|_a$ of the approximate solution $\hat{u}_n$.

Thus the problem is reduced to a problem, independent of the FEM,
of *approximating* a fixed, although unknown, function u in the energy
space and its error estimation.

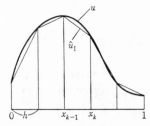

Fig. 3.3 Piecewise linear interpolate $\hat{u}_1$ of u.

Usually an *interpolate* $\hat{u}_1$ of u is chosen for $\hat{v}_n$ as a close approximation to u. The subscript I represents interpolation.

Consider again problem (2.3.9). We assume that the boundary condition here may be of the inhomogeneous Dirichlet type. In exactly the same way as in the FEM we divide the interval $[0,1]$ into n equal subintervals by the nodes

$$x_k = kh \qquad k = 0,1,\ldots,n \qquad (3.7.1)$$

and define a continuous piecewise linear polynomial $\hat{u}_1(x)$ satisfying

$$\hat{u}_1(x_k) = u(x_k) \qquad k = 0,1,\ldots,n \qquad (3.7.2)$$

at each node, as shown in Fig. 3.3. This is the simplest interpolate of u based on a piecewise polynomial. It is obvious that, since the nodes are fixed, $\hat{u}_1$ is determined uniquely by u.

The interpolate $\hat{u}_1$ belongs to $H = H_1$ and satisfies the given boundary condition. Therefore we can choose $\hat{u}_1$ as a trial function in the problem, that is, as $\hat{v}_n$ in (3.4.8).

3.8 *Error of the Piecewise Linear Polynomial*

Since we have chosen $\hat{u}_1$ as $\hat{v}_n$ in the inequality (3.4.8), next we have to estimate its error. In order to do this we consider here the error

$$e(x) = u(x) - \hat{u}_1(x) \qquad (3.8.1)$$

of the interpolate $\hat{u}_1$ of u, where u is a given function. Since $\hat{u}_1$ agrees with u at each node, the difference e vanishes at each node. Therefore, if we assume that u'' is square-integrable over $(0,1)$, we have from (2.5.10)

$$\int_{xk}^{xk+1} e^2 \, dx \le \frac{h^4}{\pi^4} \int_{xk}^{xk+1} e''^2 \, dx = \frac{h^4}{\pi^4} \int_{xk}^{xk+1} u''^2 \, dx \qquad (3.8.2)$$

The last equality follows from the fact that $\hat{u}_1'' = 0$ over the subinterval (x_k, x_{k+1}) because $\hat{u}_1$ is piecewise linear. Summing up (3.8.2) over the entire interval we have

$$\int_0^1 e(x)^2 \, dx \le \frac{h^4}{\pi^4} \int_0^1 u''^2 \, dx \tag{3.8.3}$$

This inequality shows that, as long as the second derivative of u is square-integrable, the mean square error of the interpolate $\hat{u}_1$ given by the square root of the left-hand side is of order h^2.

As above, we have from (2.5.9)

$$\int_0^1 e'(x)^2 \, dx \le \frac{h^2}{\pi^2} \int_0^1 u''^2 \, dx \tag{3.8.4}$$

Then from (2.1.5) we obtain

$$a(e,e) \le M \int_0^1 (e'^2 + e^2) \, dx \le M \left(\frac{h^2}{\pi^2} + \frac{h^4}{\pi^4} \right) \int_0^1 u''^2 \, dx \tag{3.8.5}$$

which gives, from the definition of the energy norm (3.3.1),

$$\| u - \hat{u}_1 \|_a \le Ch \left\{ \int_0^1 u''^2 \, dx \right\}^{1/2} \tag{3.8.6}$$

where C is a constant that does not depend on h if h is smaller than some fixed h_0. Thus we see that, when h is sufficiently small, the error of the interpolate is of order h in the energy norm.

3.9 Estimation of the Second Derivative of the Solution

Again we go back to the boundary value problem (1.1.1) and (1.1.2). The right-hand side of (3.8.6) includes the second derivative of the solution u of the problem. If we can further estimate the right-hand side in terms of the known function f that appears on the right-hand side of the given equation, evaluation of the estimate will become more specific. As already seen in Sec. 2.7, the smoothness or the differentiability of f transfers directly to u'', so that it is naturally expected that if f is small then u'' is also small. In fact, in our problem an inequality

$$\int_0^1 u''^2 \, dx \le C \int_0^1 f^2 \, dx \tag{3.9.1}$$

where C is a constant holds provided that p, q, and f each satisfies an appropriate condition.

For simplicity we assume here that p and q are constant and show that for problem (1.1.1) and (1.1.2) the inequality (3.9.1) actually holds. First we expand the function $f(x)$ on the right-hand side in terms of a Fourier series:

$$f(x) = \frac{1}{2} b_0 + \sum_{j=1}^{\infty} b_j \cos j\pi x + \sum_{j=1}^{\infty} c_j \sin j\pi x \qquad (3.9.2)$$

$$b_j = 2 \int_0^1 f(x) \cos j\pi x \, dx \qquad c_j = 2 \int_0^1 f(x) \sin j\pi x \, dx \quad (3.9.3)$$

This immediately leads to *Parceval's identity*

$$\frac{1}{4} b_0^2 + \frac{1}{2} \sum_{j=1}^{\infty} b_j^2 + \frac{1}{2} \sum_{j=1}^{\infty} c_j^2 = \int_0^1 f(x)^2 \, dx \qquad (3.9.4)$$

On the other hand, if we expand the solution $u(x)$ as

$$u(x) = \sum_{j=1}^{\infty} a_j \sin j\pi x \qquad (3.9.5)$$

the coefficients $\{a_j\}$ are given by (1.1.6). Therefore we have

$$u''(x) = -\sum_{j=1}^{\infty} (j\pi)^2 a_j \sin j\pi x$$

$$= -\sum_{j=1}^{\infty} \frac{(j\pi)^2}{(j\pi)^2 p + q} c_j \sin j\pi x \qquad (3.9.6)$$

If we integrate the square of both sides of (3.9.6), we obtain from the orthogonality of $\{\sin j\pi x\}$ and from (3.9.4) the following inequality which we sought:

$$\int_0^1 u''(x)^2 \, dx = \frac{1}{2} \sum_{j=1}^{\infty} \left\{ \frac{(j\pi)^2}{(j\pi)^2 p + q} \right\}^2 c_j^2 \le \frac{1}{2p^2} \sum_{j=1}^{\infty} c_j^2$$

$$\le \frac{1}{p^2} \int_0^1 f(x)^2 \, dx \qquad (3.9.7)$$

Once the estimate (3.9.1) is established, we have, as an explicit estimate of the error of the interpolate,

$$\|u - \hat{u}_1\|_a \le C'h \left\{ \int_0^1 f^2 \, dx \right\}^{1/2} \qquad (3.9.8)$$

from (3.8.6). Hereafter we assume that for the second derivative of the solution an inequality like (3.9.1) always holds.

3.10 Estimation in Terms of the Norm of a Continuous Linear Functional

The inequality (3.9.1) gives an estimate of the second derivative of the solution u, and we conclude from it that if the norm of f changes continuously then the norm of u'' also changes continuously. That is, this inequality represents a continuity in this sense between f and u'' appearing in (1.1.1) or (2.4.4) that includes the second derivative explicitly.

We can also obtain an expression that represents a similar continuity relation for the weak form (2.3.8). Before we present it we comment on the integral

$$(f, u) = \int_0^1 fu \, dx \tag{3.10.1}$$

appearing in (2.3.8) and (2.1.9). Note that this is not an inner product, although it has the same form as an inner product, because f and u do not belong to the same function space. For example, in (2.7.1) the right-hand side is $f = \delta(x - \frac{1}{2}) = \delta_{1/2}$, while we think of u in (3.10.1) multiplied by f as an element of $\mathring{H}_1$. Also, in this case the integral (3.10.1) makes sense and gives $u(\frac{1}{2})$.

Thus we denote by H_{-1} the space consisting of all functions f for which the integral (3.10.1) exists for any $u \in \mathring{H}_1$. In other words, H_{-1} is a space consisting of all continuous linear functionals over $\mathring{H}_1$, and the Dirac δ-function $\delta(x - \frac{1}{2}) = \delta_{1/2}$ is an element. We define the norm of H_{-1} as

$$\|f\|_{-1} = \sup_{v \in \mathring{H}_1} \frac{|(f, v)|}{\|v\|_1} \tag{3.10.2}$$

Consider $f = \delta_{1/2}$ again as an example. In the Schwarz inequality (3.1.14) we let $a = 0$ and $b = \frac{1}{2}$ and replace u and v by 1 and v', respectively. Then, from $v(0) = 0$ we have

$$\left\{ v\left(\frac{1}{2}\right) \right\}^2 \le \frac{1}{2} \int_0^{1/2} v'^2 \, dx \le \int_0^1 v'^2 \, dx \le \|v\|_1^2 \tag{3.10.3}$$

This leads to

$$\frac{|(\delta_{1/2}, v)|}{\|v\|_1} = \frac{|v(\frac{1}{2})|}{\|v\|_1} \le 1 \qquad \forall v \in \mathring{H}_1 \tag{3.10.4}$$

Thus we see that

$$\|\delta_{1/2}\|_{-1} \le 1 \tag{3.10.5}$$

That is, the H_{-1} norm of the δ-function is bounded.

Returning to the solution u of (2.3.8), we put $\eta = u$ in (2.3.8). Then we have

$$a(u, u) = (f, u) \tag{3.10.6}$$

the left-hand side of which is bounded from below as

$$\gamma \|u\|_1^2 \le a(u, u) \tag{3.10.7}$$

by the ellipticity of (2.5.13). On the other hand, the absolute value of the right-hand side of (3.10.6) is bounded from above by (3.10.2) as

$$|(f, u)| \leq \|f\|_{-1} \|u\|_1 \tag{3.10.8}$$

Therefore, from these inequalities divided by $\gamma \|u\|_1$ we obtain

$$\|u\|_1 \leq \frac{1}{\gamma} \|f\|_{-1} \tag{3.10.9}$$

This inequality implies that, even in the case where f on the right-hand side of the equation is the Dirac δ-function, if its deviation is small for example, if the deviation of the position of the point load is small, the solution u together with its first derivative never becomes too large.

3.11 *Error Estimation of the Finite Element Solution in Terms of the Energy Norm*

Since the error of the interpolate $\hat{u}_I$ is bounded as (3.8.6), we next substitute $\hat{u}_I$ for $\hat{v}_n$ in the inequality (3.4.8), which leads to an error estimate of the finite element solution in terms of the energy norm:

$$\|u - \hat{u}_n\|_a \leq Ch \left\{ \int_0^1 u''^2 \, dx \right\}^{1/2} \tag{3.11.1}$$

This implies that, as far as the second derivative u'' of the exact solution is square-integrable, the error in the energy norm is of order h. Although the error of the interpolate $\hat{u}_I$ is of order h, from (3.8.3), the order in h is decreased by 1 because the energy norm includes the first derivative which is of order h as seen from (3.8.4).

Since the bilinear form $a(u,v)$ is positive definite, we have an estimate of the error in terms of the Sobolev norm

$$\|u - \hat{u}_n\|_1 \leq C'h \left\{ \int_0^1 u''^2 \, dx \right\}^{1/2} \tag{3.11.2}$$

from (2.5.12). This indicates not only that the finite element solution $\hat{u}_n$ itself approaches the exact solution u but also that the first derivative $\hat{u}'_n$ approaches u' as h tends to zero.

3.12 *Mean Square Error of the Finite Element Solution and Nitsche's Trick*

In this section we derive an estimate of the error of the finite element solution $\hat{u}_n$ itself, that is, a mean square estimate of the error without a derivative term. It has already been seen in (3.8.3) that the mean square error of the interpolate $\hat{u}_1$ satisfies

$$\|u - \hat{u}_1\|_0 \equiv \left\{ \int_0^1 (u - \hat{u}_1)^2 \, dx \right\}^{1/2} \le \frac{h^2}{\pi^2} \left\{ \int_0^1 u''^2 \, dx \right\}^{1/2} \quad (3.12.1)$$

The subscript 0 indicates the zeroth derivative, that is, the Sobolev norm that includes only the function value. Although the finite element solution $\hat{u}_n$ corresponding to the exact solution u is not equal to the interpolate $\hat{u}_1$ of u, these two values are perhaps close to each other and the mean square error of $\hat{u}_n$ is also expected to be of order h^2.

A skillful method for deriving an estimate of the mean square error is known. It is called *Nitsche's trick* and is as follows. Denote the error of the finite element solution $\hat{u}_n$ by

$$\hat{e}_n = u - \hat{u}_n \quad (3.12.2)$$

Consider an auxiliary problem that has $\hat{e}_n$ instead of f on the right-hand side of (2.3.8):

$$a(w, \eta) = (\hat{e}_n, \eta) \qquad \forall \eta \in \overset{\circ}{H}_1 \quad (3.12.3)$$

We assume that the solution w satisfies $w = 0$ on the boundary. Notice that u and $\hat{u}_n$ coincide with each other on the boundary, hence $\hat{e}_n$ belongs to the space $\overset{\circ}{H}_1$. If $\eta = \hat{e}_n$, (3.12.3) becomes

$$a(w, \hat{e}_n) = (\hat{e}_n, \hat{e}_n) = \|\hat{e}_n\|_0^2 \quad (3.12.4)$$

On the other hand, from (3.5.7)

$$a(u - \hat{u}_n, \hat{v}) = a(\hat{e}_n, \hat{v}) = a(\hat{v}, \hat{e}_n) = 0 \quad (3.12.5)$$

holds for $\hat{v} \in \overset{\circ}{K}_n$, so that we have, by subtracting (3.12.5) from (3.12.4),

$$a(w - \hat{v}, \hat{e}_n) = \|\hat{e}_n\|_0^2 \quad (3.12.6)$$

Application of the Schwarz inequality (3.3.3) to the right-hand side results in

$$\|\hat{e}_n\|_0^2 \le \{a(w - v, x - v)\}^{1/2} \{a(e_n, e_n)\}^{1/2} \quad (3.12.7)$$

We choose here a Galerkin approximation to w for $\hat{v}$. Then we have, from (3.11.1) and (3.9.7),

$$\{a(w - v, w - \hat{v})\}^{1/2} \le C'h \left\{ \int_0^1 \hat{e}_n^2 \, dx \right\}^{1/2} = C'h\|\hat{e}_n\|_0 \quad (3.12.8)$$

and also, from (3.11.1),

$$\{a(\hat{e}_n, \hat{e}_n)\}^{1/2} \le C''h \left\{ \int_0^1 u''^2 \, dx \right\}^{1/2} \quad (3.12.9)$$

Therefore, if we substitute these results into the right-hand side of

(3.12.7), we have

$$\|\hat{e}_n\|_0{}^2 \leq C'''h^2\|\hat{e}_n\|_0\left\{\int_0^1 u''^2\,dx\right\}^{1/2} \tag{3.12.10}$$

and by dividing both sides by $\|\hat{e}_n\|_0$ we eventually obtain

$$\|\hat{e}_n\|_0 \leq C'''h^2\left\{\int_0^1 u''^2\,dx\right\}^{1/2} \tag{3.12.11}$$

Thus we see that the order of the mean square error of $\hat{u}_n$ is h^2.

Note that the order h or h^2 of the error so far obtained is based on the error of the interpolate $\hat{u}_1$ of the exact solution u. If we repeat mesh refinement in the actual numerical solution of a problem with good properties we usually observe that the analysis given above is almost true. That is, in regard to the rate of convergence of the finite element solution $\hat{u}_n$ to the exact solution u when the mesh is repeatedly refined, (3.11.2) or (3.12.11) usually holds. However, it must be noted that, for a general problem in which the exact solution u is not known, the analysis given above cannot give a specific estimate of the amount of actual error for a particular division of the given domain. The reason for this is that the error estimates shown above include the unknown constants C' and C'''.

4

Elliptic Boundary Value Problems in Two Space Dimensions

4.1 *A Two-Dimensional Boundary Value Problem and the Weak Form*

In practice the FEM is applied to problems in two or three space dimensions. In this chapter we consider a boundary value problem involving a second-order elliptic partial differential equation as an example of a stationary problem in two space dimensions and apply the FEM to it.

The problem we consider here is a boundary value problem of the Dirichlet type in a bounded domain G on the xy plane:

$$-\Delta u + qu = f \qquad\qquad (4.1.1)$$

$$u = 0 \qquad \text{on } \partial G \qquad\qquad (4.1.2)$$

where ∂G represents the boundary of G and Δ is the Laplacian

$$\Delta u = \frac{\partial^2 u}{\partial x^2} + \frac{\partial^2 u}{\partial y^2} \qquad\qquad (4.1.3)$$

We assume that $q(x,y)$ and $f(x,y)$ are given functions with appropriate smoothness and that in G

$$q(x,y) \geq 0 \qquad\qquad (4.1.4)$$

In order to apply the FEM according to the principles stated so far it is first necessary to derive a weak form corresponding to (4.1.1). To this end we introduce a space of functions whose first derivative is square-integrable, that is, whose norm is defined by

$$\|v\|_1 = \left[\iint_G \left\{ \left(\frac{\partial v}{\partial x} \right)^2 + \left(\frac{\partial v}{\partial y} \right)^2 + v^2 \right\} dx \, dy \right]^{1/2} \tag{4.1.5}$$

and we denote it anew by H_1. This is a typical example of a Sobolev space. If we define the inner product by

$$(u, v)_1 = \iint_G \left(\frac{\partial u}{\partial x} \frac{\partial v}{\partial x} + \frac{\partial u}{\partial y} \frac{\partial v}{\partial y} + uv \right) dx \, dy \tag{4.1.6}$$

then H_1 becomes a Hilbert space. From the Schwarz inequality $|(u, v)_1| \leq \|u\|_1 \|v\|_1$ it can be shown that (4.1.5) satisfies condition (2.2.4) of the norm. Furthermore, we define a subspace of H_1 consisting of functions in H_1 that satisfy

$$v = 0 \qquad \text{on } \partial G \tag{4.1.7}$$

and denote it by $\mathring{H}_1$.

Since the boundary condition is of the Dirichlet type, we multiply both sides of (4.1.1) by $v \in \mathring{H}_1$, as in the preceding chapter, in order to derive a weak form of (4.1.1):

$$-\iint_G \left(\frac{\partial^2 u}{\partial x^2} + \frac{\partial^2 u}{\partial y^2} - qu \right) v \, dx \, dy = \iint_G fv \, dx \, dy \tag{4.1.8}$$

We integrate this by parts. What is necessary on this occasion is *Green's formula*:

$$\iint_G (\Delta u) v \, dx \, dy = -\iint_G \left(\frac{\partial u}{\partial x} \frac{\partial v}{\partial x} + \frac{\partial u}{\partial y} \frac{\partial v}{\partial y} \right) dx \, dy$$
$$+ \int_{\partial G} \left(\frac{\partial u}{\partial x} \cos \theta_1 + \frac{\partial u}{\partial y} \cos \theta_2 \right) v \, d\sigma$$

$$\tag{4.1.9}$$

The second term on the right-hand side is an integral along the boundary ∂G, and $\cos \theta_1$ and $\cos \theta_2$ and direction cosines in the directions of x and y, respectively, of the unit outward normal vector $\boldsymbol{n}$ to the curved element $d\sigma$ (Fig. 4.1). This term can be expressed in various forms using the direction derivative:

$$\int_{\partial G} \left(\frac{\partial u}{\partial x} \cos \theta_1 + \frac{\partial u}{\partial y} \cos \theta_2 \right) v \, d\sigma$$
$$= \int_{\partial G} (\nabla u \cdot \boldsymbol{n}) v \, d\sigma$$
$$= \int_{\partial G} \left(\frac{\partial u}{\partial x} \frac{\partial x}{\partial n} + \frac{\partial u}{\partial y} \frac{\partial y}{\partial n} \right) v \, d\sigma$$
$$= \int_{\partial G} \frac{\partial u}{\partial n} v \, d\sigma$$

$$\tag{4.1.10}$$

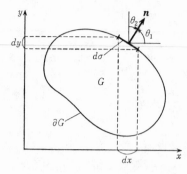

Fig. 4.1 The boundary ∂G of the domain G and the outer normal n.

where $\partial/\partial n$ represents the differentiation toward the outward normal of the boundary ∂G and ∇u is

$$\nabla u = \left(\frac{\partial u}{\partial x},\ \frac{\partial u}{\partial y}\right) \qquad (4.1.11)$$

Integrating the right-hand side of (4.1.8) by parts using (4.1.9) we have

$$\iint_G \left(\frac{\partial u}{\partial x}\frac{\partial v}{\partial x} + \frac{\partial u}{\partial y}\frac{\partial v}{\partial y} + quv - fv\right) dx\, dy - \int_{\partial G}\frac{\partial u}{\partial n}v\, d\sigma = 0$$

$$(4.1.12)$$

Note that $v \in \mathring{H}_1$, that is, $v = 0$ on the boundary ∂G. Then the second term on the left-hand side vanishes, and we obtain the weak form corresponding to (4.1.1) and (4.1.2):

$$\iint_G \left(\frac{\partial u}{\partial x}\frac{\partial v}{\partial x} + \frac{\partial u}{\partial y}\frac{\partial v}{\partial y} + quv - fv\right) dx\, dy = 0 \qquad v \in \mathring{H}_1 \qquad (4.1.13)$$

$$u = 0 \qquad \text{on } \partial G \qquad (4.1.14)$$

We rewrite this equation in a form that can be applied to general problems, as we did in the case of one space dimension. First we define a bilinear form for functions in H_1:

$$a(u,v) = \iint_G \left(\frac{\partial u}{\partial x}\frac{\partial v}{\partial x} + \frac{\partial u}{\partial y}\frac{\partial v}{\partial y} + quv\right) dx\, dy \qquad (4.1.15)$$

Next we introduce the following notation:

$$(f,v) = \iint_G fv\, dx\, dy \qquad (4.1.16)$$

If v is an element of $\mathring{H}_1$, f is an element of H_{-1} whose norm is defined by (3.10.2). Then the weak form (4.1.2) can be written as

$$a(u,v) - (f,v) = 0 \qquad \forall v \in \mathring{H}_1 \qquad (4.1.17)$$

$$u = 0 \qquad \text{on } \partial G \qquad\qquad (4.1.18)$$

4.2 Variational Formulation

The problem given in the preceding section is equivalent to the variational problem in which the functional

$$J[u] = \frac{1}{2} \iint_G \left\{ \left(\frac{\partial u}{\partial x} \right)^2 + \left(\frac{\partial u}{\partial y} \right)^2 + qu^2 - 2fu \right\} dx \, dy \qquad (4.2.1)$$

is minimized under the condition that

$$u = 0 \qquad \text{on } \partial G \qquad\qquad (4.2.2)$$

If we use the notation introduced in the preceding section, the functional $J[u]$ is written as

$$J[u] = \tfrac{1}{2} a(u, u) - (f, u) \qquad\qquad (4.2.3)$$

In order to obtain the solution of this variational problem, as in the case of one space dimension, we construct a test function

$$u_\epsilon = u + \epsilon \eta \qquad\qquad (4.2.4)$$

where $u =$ function minimizing (4.2.1)

$\epsilon =$ an arbitrary real number

$\eta =$ any function belonging to $\mathring{H}_1$

Substituting this expression for u in (4.2.1) and setting the first variation δJ equal to 0, that is, making J stationary when $\epsilon = 0$, we obtain the weak form (4.1.17).

4.3 The Ellipticity Condition

In order for the function that makes $J[u]$ stationary to actually be the function that minimizes $J[u]$, the coefficient of ϵ^2 in the second variation of (2.3.4) must be positive as seen in Sec. 2.3 or, equivalently, the bilinear form $a(u, v)$ must be elliptic as in the case of one space dimension. In order to show that $a(u, v)$ is elliptic the following inequality of the Poincaré type will be useful:

$$\frac{2\pi^2}{d^2} \iint_G v^2 \, dx \, dy \le \iint_G \left\{ \left(\frac{\partial v}{\partial x} \right)^2 + \left(\frac{\partial v}{\partial y} \right)^2 \right\} dx \, dy \qquad \forall v \in \mathring{H}_1$$

$$(4.3.1)$$

where d is the length of the diagonal of a rectangle circumscribing the domain G.

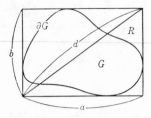

Fig. 4.2 The rectangle R circumscribed to the domain G.

This inequality is verified as follows. Consider a rectangle R circumscribing the domain G whose sides are parallel to the x and y axes and let the length of these sides be a and b, as shown in Fig. 4.2. We extend the domain of v to the entire R in such a way that $v = 0$ outside G. Then, putting $\alpha = 0$ and $\beta = a$ in (2.5.6), we have

$$\frac{\pi^2}{a^2} \int_0^a v^2 \, dx \le \int_0^1 \left(\frac{\partial v}{\partial x}\right)^2 dx \tag{4.3.2}$$

Integration with respect to y yields

$$\frac{\pi^2}{a^2} \iint_R v^2 \, dx \, dy \le \iint_R \left(\frac{\partial v}{\partial x}\right)^2 dx \, dy \tag{4.3.3}$$

Replacing x by y, we obtain

$$\frac{\pi^2}{b^2} \iint_R v^2 \, dx \, dy \le \iint_R \left(\frac{\partial v}{\partial y}\right)^2 dx \, dy \tag{4.3.4}$$

Summing (4.3.3) and (4.3.4) and taking into account that $v = 0$ outside G, we have

$$\pi^2\left(\frac{1}{a^2} + \frac{1}{b^2}\right) \iint_G v^2 \, dx \, dy \le \iint_G \left\{\left(\frac{\partial v}{\partial x}\right)^2 + \left(\frac{\partial v}{\partial y}\right)^2\right\} dx \, dy \tag{4.3.5}$$

Finally, the inequality

$$\frac{1}{a^2} + \frac{1}{b^2} \ge \frac{2}{a^2 + b^2} = \frac{2}{d^2} \tag{4.3.6}$$

leads to (4.3.1).

The bilinear form $a(u,v)$ of the problem is defined by (4.1.15), so that from $q \ge 0$ we have

$$a(v,v) \ge \iint_G \left\{\left(\frac{\partial v}{\partial x}\right)^2 + \left(\frac{\partial v}{\partial y}\right)^2\right\} dx \, dy$$

$$= \frac{1}{1 + (d^2/2\pi^2)} \iint_G \left\{\left(\frac{\partial v}{\partial x}\right)^2 + \left(\frac{\partial v}{\partial y}\right)^2\right\}\left(1 + \frac{d^2}{2\pi^2}\right) dx \, dy \tag{4.3.7}$$

which, together with (4.3.1), leads to the *ellipticity condition* for $a(u,v)$

$$\gamma\|v\|_1^2 \le a(v,v) \qquad \forall v \in \mathring{H}_1 \qquad (4.3.8)$$

where

$$\gamma = \frac{1}{1 + (d^2/2\pi^2)} \qquad (4.3.9)$$

Thus the coefficient of ϵ^2 in the term corresponding to the second variation of (2.3.4) is positive, and we conclude that $J[u_\epsilon]$ actually attains its minimum at the solution u.

4.4 *Triangulation of the Domain and the Basis Functions*

Since the weak form (4.1.13) has been obtained, next we need to construct basis functions appropriate to the FEM for two space dimensions.

When we apply the FEM to a problem involving a domain with a curved boundary, we usually need to approximate the shape of the domain by, say, a polygon. In this chapter, however, we assume from the beginning that the given domain G consists of a polygon so that we do not need to approximate its shape.

We divide the given polygonal domain G into small triangles as shown in Fig. 4.3. The vertices on the boundary are chosen in such a way that each of them coincides with the vertices of some small triangles. We also assume that, although a vertex of a triangle may be located on the boundary, one is never located on the side of another triangle. The small triangle is called the *triangular element* of the FEM, and the vertex of the triangle is called the *node*. The nodes should be numbered with care. As will be seen later, it is convenient from the viewpoint of numerical computation that adjacent nodes have numbers as close as possible to each other.

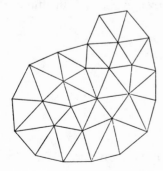

Fig. 4.3 Triangulation of the polygonal domain G.

Triangulation is not the only way to subdivide the domain. For example, when the shape of the given domain is a rectangle, it is possible to divide it into small rectangles and to construct basis functions based on these rectangles. However, although there is a difference between this kind of subdivision and triangulation from a technical point of view, there is no essential difference from a standpoint of principle, hence in this book we will restrict ourselves to triangulation.

4.5 Basis Functions in Two Space Dimensions and the Finite Element Solution

We define a pyramid-shaped function $\hat{\phi}_k(x, y)$ such that it takes the value 1 at the kth node while it vanishes at other nodes, and that in each triangular element it is a plane (Fig. 4.4). That is, $\hat{\phi}_k(x, y)$ is a piecewise linear function satisfying

$$\hat{\varphi}_k(x_j, y_j) = \begin{cases} 1 & j = k \\ 0 & j \neq k \end{cases} \tag{4.5.1}$$

where (x_j, y_j) are the coordinates of the jth node. This is the simplest and most important basis function used for two-dimensional problems. In order to express a boundary value that does not vanish, we construct a basis function such that it takes the value 1 only at a particular node on the boundary, vanishes at other nodes, and is identically 0 outside the domain.

Let the number of nodes that are not on the boundary, that is, the number of interior points, be n. Then

$$\hat{u}_n(x, y) = \sum_{j=1}^{n} a_j \hat{\varphi}_j(x, y) \tag{4.5.2}$$

where the summation is taken with respect to the interior nodes, serves as a trial function for (4.1.13) and (4.1.14). As shown in Fig. 4.5, this func-

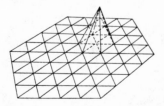

Fig. 4.4 Piecewise linear basis function $\hat{\phi}_k(x, y)$.

Fig. 4.5 Piecewise linear trial function $\hat{u}_n(x,y)$.

tion has a shape such that it vanishes at the boundary nodes, is linear and continuous along the sides of the triangular elements, and is a plane in each triangular element. As in the case of one space dimension, the value of the coefficient a_j is equal to the value of $\hat{u}_n(x,y)$ at the jth node, that is,

$$a_j = \hat{u}_n(x_j, y_j) \tag{4.5.3}$$

This is very convenient when we consider the physical meaning of $\hat{u}_n(x,y)$.

Next we fix the nodes, and then we denote by $\mathring{K}_n$ the space of all functions given by (4.5.2). It is obvious that it is a subspace of $\mathring{H}_1$.

When searching for an approximate solution to (4.1.13), we treat the problem in the n-dimensional subspace $\mathring{K}_n$, so that it is sufficient to take n independent $\hat{\phi}_j(x,y)$, $j = 1,2,\ldots,n$, for v in (4.1.13). Thus (4.1.13) is replaced by

$$\iint_G \left(\frac{\partial \hat{u}_n}{\partial x} \frac{\partial \hat{\phi}_j}{\partial x} + \frac{\partial \hat{u}_n}{\partial y} \frac{\partial \hat{\phi}_j}{\partial y} + q\hat{u}_n\hat{\phi}_j - f\hat{\phi}_j \right) dx\,dy = 0 \qquad j = 1,2,\ldots,n$$
$$\tag{4.5.4}$$

$$\hat{u}_n = 0 \qquad \text{on } \partial G \tag{4.5.5}$$

Using (4.1.17) this can also be written as

$$a(\hat{u}_n, \hat{\phi}_j) - (f, \hat{\phi}_j) = 0 \qquad j = 1,2,\ldots,n \tag{4.5.6}$$

$$\hat{u}_n = 0 \qquad \text{on } \partial G \tag{4.5.7}$$

Substituting (4.5.2) into these equations, we have a system of linear equations

$$(K + M)\mathbf{a} = \mathbf{f} \tag{4.5.8}$$

where $\mathbf{a}$ is an n-dimensional vector whose jth entry is a_j, $\mathbf{f}$ is an n-dimensional vector whose jth entry is

$$f_j = \iint_G f(x,y)\hat{\phi}_j(x,y)\,dx\,dy \tag{4.5.9}$$

and K and M are matrices whose ij entries are

$$K_{ij} = \iint_G \left(\frac{\partial \hat{\varphi}_i}{\partial x} \frac{\partial \hat{\varphi}_j}{\partial x} + \frac{\partial \hat{\varphi}_i}{\partial y} \frac{\partial \hat{\varphi}_j}{\partial y} \right) dx \, dy \qquad (4.5.10)$$

and

$$M_{ij} = \iint_G q \hat{\varphi}_i \hat{\varphi}_j \, dx \, dy \qquad (4.5.11)$$

respectively. As in the case of one space dimension, K and M are called the stiffness matrix and the mass matrix, respectively.

The matrices K and M are symmetric. Let b be any vector whose jth entry is b_j. If $b \neq 0$, then

$$b^T K b = \iint_G \left\{ \left(\sum_{j=1}^n b_j \frac{\partial \hat{\varphi}_j}{\partial x} \right)^2 + \left(\sum_{j=1}^n b_j \frac{\partial \varphi_j}{\partial y} \right)^2 \right\} dx \, dy > 0$$

$$(4.5.12)$$

holds, hence K is positive definite. Furthermore, from $q \geq 0$, M satisfies $b^T M b \geq 0$. Thus $K + M$ is positive definite and (4.5.8) has a unique solution, so that substitution of solution a_j into (4.5.2) leads to a finite element solution $\hat{u}_n$.

Since the basis functions $\hat{\phi}_j(x,y)$ are identically zero except in a small number of subdomains, where we say that $\hat{\phi}_j(x,y)$ has *local support*, most of the entries of $K + M$ vanish; that is, $K + M$ is sparse. Distribution of the nonzero entries of $K + M$, however, depends largely on how the triangular elements are numbered. It is evident from the definition of the basis function $\hat{\phi}_j$ that the entry $(K + M)_{jk}$ of $K + M$ vanishes unless the jth and kth nodes are identical or adjacent to each other. Hence by numbering the adjacent nodes as closely as possible to each other we can locate the nonzero entries in the neighborhood of the diagonal of the matrix. One of the standard methods for solving a system of linear equation is the *Gauss elimination method*. When applying the Gauss elimination method it is desirable that the nonzero entries be concentrated in the neighborhood of the diagonal from the standpoint of efficient computation and the saving of computer memory. Hence we should number the nodes with discretion.

4.6 The Natural Boundary Condition and the Mixed Boundary Condition

The boundary conditions considered so far have been of the homogeneous Dirichlet type. In this section we will deal with a problem involving G with a little more complicated boundary condition on ∂G as follows.

We divide the boundary ∂G into two portions, ∂G_1 and ∂G_2:

$$-\Delta u + qu = f \qquad (4.6.1)$$

$$u = 0 \qquad \text{on } \partial G_1 \qquad (4.6.2)$$

$$\frac{\partial u}{\partial n} + \alpha(x,y)u + \beta(x,y) = 0 \qquad \text{on } \partial G_2 \qquad (4.6.3)$$

Again we assume $q \geq 0$, and $\partial/\partial n$ denotes differentiation toward the outward normal along the boundary ∂G_2. Condition (4.6.2) is of the Dirichlet type, and (4.6.3) is a natural boundary condition in the following sense.

In this section, by H_1^* we mean the space of functions whose norm is defined by

$$\|v\|_1 = \iint_G \left\{ \left(\frac{\partial v}{\partial x}\right)^2 + \left(\frac{\partial v}{\partial y}\right)^2 + v^2 \right\} dx\, dy \qquad (4.6.4)$$

such that they vanish on the boundary ∂G_1. Nothing is assumed on the boundary ∂G_2. Multiplying both sides of (4.6.1) by $v \ni H_1^*$ and integrating by parts over G using Green's formula (4.1.9) and the boundary conditions (4.6.2) and (4.6.3), we obtain a weak form corresponding to (4.6.1) through (4.6.3):

$$\iint_G \left(\frac{\partial u}{\partial x}\frac{\partial v}{\partial x} + \frac{\partial u}{\partial y}\frac{\partial v}{\partial y} + quv - fv \right) dx\, dy$$

$$+ \int_{\partial G_2} (\alpha u + \beta) v\, d\sigma = 0 \qquad \forall v \in H_1^* \qquad (4.6.5)$$

$$u = 0 \qquad \text{on } \partial G_1 \qquad (4.6.6)$$

We define the bilinear form $a(u,v)$ for $u,v \in H_1^*$ and the boundary integral $\langle \alpha u + \beta, v \rangle$ by

$$a(u,v) = \iint_G \left(\frac{\partial u}{\partial x}\frac{\partial v}{\partial x} + \frac{\partial u}{\partial y}\frac{\partial v}{\partial y} + quv \right) dx\, dy \qquad (4.6.7)$$

$$\langle \alpha u + \beta, v \rangle = \int_{\partial G_2} (\alpha u + \beta) v\, d\sigma$$

and rewrite (4.6.5) and (4.6.6) as

$$a(u,v) + \langle \alpha u + \beta, v \rangle - (f,v) = 0 \qquad \forall v \in H_1^* \qquad (4.6.8)$$

$$u = 0 \qquad \text{on } \partial G_1 \qquad (4.6.9)$$

Integrating the weak form (4.6.5) by parts toward the original equation using (4.6.2), we have

$$\iint_G (-\Delta u + qu - f) v \, dx \, dy + \int_{G_2} \left(\frac{\partial u}{\partial n} + \alpha u + \beta\right) v \, d\sigma = 0$$

(4.6.10)

First we choose an arbitrary v satisfying $v = 0$ on ∂G_2 as $v \in H_1{}^*$, obtaining (4.6.1), and the first term of the right-hand side of (4.6.10) vanishes. Next the condition that the second term of the right-hand side of (4.6.10) vanishes for any $v \in H_1{}^*$ whose boundary value on ∂G_2 is arbitrary gives (4.6.3). That is, (4.6.3) is a natural boundary condition.

It is easy to see that the boundary value problem given above is equivalent to the variational problem in which the following functional $J[u]$ is minimized:

$$J[u] = \tfrac{1}{2} a(u, u) + \langle \alpha u + \beta, u \rangle - (f, u) \qquad (4.6.11)$$

$$u = 0 \qquad \text{on } \partial G_1 \qquad (4.6.12)$$

The condition (4.6.2) and (4.6.3), which includes both the Dirichlet condition and the natural condition, is called a *mixed-type boundary condition*.

The natural boundary condition is satisfied automatically by the solution of (4.6.5) or by the minimizing function of (4.6.11). Therefore, as already pointed out, when applying the FEM to (4.6.5) the condition (4.6.3) on ∂G_2 need not be assumed for the trial function $\hat{u}_n$, while the condition (4.6.2) on ∂G_1 must by imposed on $\hat{u}_n$. For example, consider a problem in which (4.6.1) through (4.6.3) are solved for the domain given by Fig. 4.6. In this case we construct a trial function (4.5.2) by choosing the values at the n $(=15)$ nodes with small black circles, including those on the boundary ∂G_2, as n unknowns $a_j, j = 1, 2, \ldots, n$. Then we substitute this function into the approximate equation

$$\iint_G \left(\frac{\partial \hat{u}_n}{\partial x} \frac{\partial \hat{\varphi}_k}{\partial x} + \frac{\partial \hat{u}_n}{\partial y} \frac{\partial \hat{\varphi}_k}{\partial y} + q \hat{u}_n \hat{\varphi}_k\right) dx \, dy + \int_{\partial G_2} \alpha \hat{u}_n \hat{\varphi}_k \, d\sigma$$
$$= \iint_G f \hat{\varphi}_k \, dx \, dy - \int_{\partial G_2} \beta \hat{\varphi}_k \, d\sigma \qquad k = 1, 2, \ldots, n$$

(4.6.13)

corresponding to (4.6.5). The basis functions corresponding to the nodes on ∂G_2 are considered to vanish outside the domain. Substituting (4.5.2) tor $\hat{u}_n$ we have a system of linear equations with respect to $a_1, a_2, \ldots, a_n$:

$$\sum_{j=1}^n \left(K_{kj} + M_{kj} + \int_{\partial G_2} \alpha \hat{\varphi}_k \hat{\varphi}_j \, d\sigma\right) a_j = f_k - \int_{\partial G_2} \beta \hat{\varphi}_k \, d\sigma \qquad (4.6.14)$$

where K_{kj}, M_{kj}, and f_k are the entries of the matrices and the vector defined by (4.5.10), (4.5.11), and (4.5.9), respectively. We eventually ob-

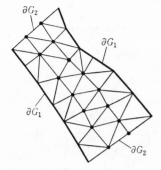

Fig. 4.6 The boundaries ∂G_1 and ∂G_2.

tain the finite element solution $\hat{u}_n$ by solving this system of equations for $\{a_j\}$.

4.7 *The Inhomogeneous Dirichlet Boundary Condition*

In this section we consider a problem involving an inhomogeneous Dirichlet boundary condition

$$-\Delta u + qu = f \qquad\qquad (4.7.1)$$

$$u = g(x, y) \qquad \text{on } \partial G \qquad\qquad (4.7.2)$$

There are several ways to treat this type of boundary condition. In the first method we consider in this section we expand the trial function in terms of the basis functions corresponding to all the nodes including those on the boundary:

$$\hat{u}_n(x, y) = \sum_{j=1}^{n+\nu} a_j \hat{\varphi}_j(x, y) \qquad\qquad (4.7.3)$$

where the interior nodes are numbered, for simplicity, from 1 to n and the boundary nodes from $n + 1$ to $n + \nu$. Then we write

$$\sum_{j=1}^{n+\nu} a_j \hat{\varphi}_j(x_l, y_l) = g(x_l, y_l) \qquad l = n + 1, \ldots, n + \nu \qquad (4.7.4)$$

that is,

$$a_l = g_l \equiv g(x_l, y_l) \qquad\qquad (4.7.5)$$

so that the function $\hat{u}_n(x, y)$ satisfies (4.7.2) at every boundary node for $P(x_l, y_l)$, $l = n + 1, \ldots, n + \nu$. This leads to

$$\hat{u}_n(x, y) = \sum_{j=1}^{n} a_j \hat{\varphi}_j(x, y) + \sum_{j=n+1}^{n+\nu} g_j \hat{\varphi}_j(x, y) \qquad (4.7.6)$$

Substituting $\hat{u}_n$ for u in (4.7.1), multiplying by $\hat{\phi}_k$, $k = 1, 2, \ldots, n$, corresponding to the interior nodes, and integrating by parts, we have

$$\sum_{j=1}^{n} a_j \iint_G \left(\frac{\partial \hat{\varphi}_j}{\partial x} \frac{\partial \hat{\varphi}_k}{\partial x} + \frac{\partial \hat{\varphi}_j}{\partial y} \frac{\partial \hat{\varphi}_k}{\partial y} + q \hat{\varphi}_j \hat{\varphi}_k \right) dx \, dy$$

$$+ \sum_{j=n+1}^{n+v} g_j \iint_G \left(\frac{\partial \hat{\varphi}_j}{\partial x} \frac{\partial \hat{\varphi}_k}{\partial x} + \frac{\partial \hat{\varphi}_j}{\partial y} \frac{\partial \hat{\varphi}_k}{\partial y} + q \hat{\varphi}_j \hat{\varphi}_k \right) dx \, dy = \iint_G f \int_k dx \, dy$$

$$k = 1, 2, \ldots, n \tag{4.7.7}$$

The boundary integral appearing when integrating by parts vanishes because $\hat{\phi}_k$, $k = 1, 2, \ldots, n$, corresponding to the interior nodes, vanishes on the boundary ∂G. Moving the terms with known boundary values to the right-hand side, we obtain the following system of n linear equations with respect to the unknowns $a_1, a_2, \ldots, a_n$:

$$\sum_{j=1}^{n} (K_{kj} + M_{kj}) a_j = - \sum_{j=n+1}^{n+v} (K_{kj} + M_{kj}) g_j + f_k \qquad k = 1, 2, \ldots, n$$

$$\tag{4.7.8}$$

where K_{kj}, M_{kj}, and f_k are the entries of the matrices (4.5.10) and (4.5.11) and the vector (4.5.9), respectively. Its solution $\{a_j\}$ gives a finite element solution $\hat{u}_n$.

Although $\hat{u}_n$ thus obtained agrees with g at each boundary node, it does not always agree with g along the boundary ∂G because $\hat{u}_n$ is a linear function while in general g is not. Therefore it must be noted that $\hat{u}_n$ is not an admissible function of the problem because (4.7.6) does not rigorously satisfy the given boundary condition.

4.8   The Penalty Method

The second way to deal with an inhomogeneous Dirichlet boundary condition is to incorporate the square integral of the boundary condition along with an unknown multiplier into the functional and to minimize it, which can be regarded as a method analogous to the Lagrange method of multipliers. Let λ be a positive constant and introduce the functional

$$J_\lambda[u] = \frac{1}{2} \iint_G \left\{ \left(\frac{\partial u}{\partial x} \right)^2 + \left(\frac{\partial u}{\partial y} \right)^2 + qu^2 - 2fu \right\} dx \, dy + \frac{1}{2} \lambda \, D[u]$$

$$\tag{4.8.1}$$

$$D[u] = \int_{\partial G} (u - g)^2 \, d\sigma \tag{4.8.2}$$

The space of the admissible functions here is H_1, introduced in Sec. 4.1. Although λ is the parameter corresponding to the Lagrange multiplier, here it is fixed as a constant. The Lagrange method of multipliers will be discussed again in Chap. 14.

Let the variation of u be δu. We do not assume $\delta u = 0$ on the boundary ∂G. Then

$$J_\lambda[u + \delta u] - J_\lambda[u]$$
$$= \iint_G \left(\frac{\partial u}{\partial x} \frac{\partial \delta u}{\partial x} + \frac{\partial u}{\partial y} \frac{\partial \delta u}{\partial y} + qu \, \delta u - f \, \delta u \right) dx \, dy$$
$$+ \lambda \int_{\partial G} (u - g) \, \delta u \, d\sigma$$
$$+ \frac{1}{2} \iint_G \left[\left(\frac{\partial \delta u}{\partial x} \right)^2 + \left(\frac{\partial \delta u}{\partial y} \right)^2 + q(\delta u)^2 \right] dx \, dy$$
$$+ \frac{1}{2} \lambda \int_{\partial G} (\delta u)^2 \, d\sigma \tag{4.8.3}$$

If we assume that u has a second-order derivative, we have from (4.1.9) and (4.1.10)

$$J_\lambda[u + \delta u] - J_\lambda[u] = \iint_G (-\Delta u + qu - f) \, \delta u \, dx \, dy$$
$$+ \lambda \int_{\partial G} \left(u - g + \frac{1}{\lambda} \frac{\partial u}{\partial n} \right) \delta u \, d\sigma$$
$$+ \frac{1}{2} \iint_G \left[\left(\frac{\partial \delta u}{\partial x} \right)^2 + \left(\frac{\partial \delta u}{\partial y} \right)^2 + q(\delta u)^2 \right] dx \, dy + \frac{1}{2} \lambda \int_{\partial G} (\delta u)^2 \, d\sigma$$

$$\tag{4.8.4}$$

For u that minimizes $J_\lambda[u]$ the first variation must vanish:

$$\delta J_\lambda[u] = \iint_G (-\Delta u + qu - f) \, \delta u \, dx \, dy$$
$$+ \lambda \int_{\partial G} \left(u - g + \frac{1}{\lambda} \frac{\partial u}{\partial n} \right) \delta u \, d\sigma = 0$$

$$\tag{4.8.5}$$

For δu we first choose any variation satisfying $\delta u = 0$ on ∂G, which leads to (4.7.1). Next we choose any variation δu on which nothing is imposed on ∂G, which leads to the natural boundary condition

$$u + \frac{1}{\lambda} \frac{\partial u}{\partial n} = g \qquad \text{on } \partial G \tag{4.8.6}$$

Note that as λ becomes large (4.8.6) approaches the given boundary condition (4.7.2).

Thus we fix λ at an appropriately large value and minimize $J_\lambda[u]$ for

$u \in H_1$. Then we obtain a solution that approximately satisfies the boundary condition (4.7.2). Since the quadratic term, that is, the second variation, in the difference of the functional (4.8.3) or (4.8.4) is positive in the neighborhood of the solution, the solution really gives the minimum of $J_\lambda[u]$. Since (4.8.6) is a natural boundary condition, it is not necessary to impose it on the trial function in the present variational problem.

In actual computation we solve the equation in the weak form

$$\iint_G \left(\frac{\partial u}{\partial x} \frac{\partial v}{\partial x} + \frac{\partial u}{\partial y} \frac{\partial v}{\partial y} + quv - fv \right) dx\, dy$$
$$+ \lambda \int_{\partial G} (u - g) v\, d\sigma = 0 \qquad v \in H_1$$

$$(4.8.7)$$

corresponding to the first variation $\delta J_\lambda = 0$ of (4.8.3) instead of minimizing $J_\lambda[u]$. More specifically, we fix λ at an appropriately large value and construct a trial function (4.7.3) in terms of the basis functions corresponding to all the nodes including those on the boundary. Then we substitute this function into the weak form

$$\iint_G \left(\frac{\partial \hat{u}_n}{\partial x} \frac{\partial \hat{\varphi}_j}{\partial x} + \frac{\partial \hat{u}_n}{\partial y} \frac{\partial \hat{\varphi}_j}{\partial y} + q\hat{u}_n\hat{\varphi}_j \right) dx\, dy + \lambda \int_{\partial G}^{1} \hat{u}_n\hat{\varphi}_j\, d\sigma$$
$$= \iint_G f\hat{\varphi}_j\, dx\, dy + \lambda \int_{\partial G} g\hat{\varphi}_j\, d\sigma \qquad j = 1, 2, \ldots, n + \nu$$

$$(4.8.8)$$

which is equivalent to (4.8.7). We solve the system of $n + \nu$ linear equations thus obtained for $\{a_j\}$, the coefficients of the finite element solution (4.7.3).

We choose a large positive value for λ in the functional (4.8.1). Then the operation to minimize $J_\lambda[u]$ gives rise to a reduction in $\lambda D[u]$, that is, a reduction in $D[u]$ itself, and as a result makes u close to g on the boundary. In other words, if a penalty $\lambda D[u]$ is imposed on the functional, the minimization of $J_\lambda[u]$ acts to reduce the penalty. The minimization method in which an integral of the given constraint multiplied by a large λ is incorporated into the functional is called the *penalty method*.

Not only for the boundary condition but also for other general constraints, the penalty method in which the constraint is incorporated into the functional usually produces an effective result. However, note that in general problems the second variation is not always positive in the neighborhood of the function that makes $J_\lambda[u]$ stationary, so that in many cases $J_\lambda[u]$ is not a minimum there. Hence it should be noted that the method of error estimation based on minimization of the functional described in the preceding chapter does not apply in such cases.

4.9   The Analytic Inhomogeneous Dirichlet Boundary Condition

Suppose the function $g(x,y)$ that prescribes the boundary condition (4.7.2) is given as an analytic function of x and y. Then we choose an analytic function that is differentiable in G such that

$$w(x,y) = g(x,y) \qquad \text{on } \partial G \tag{4.9.1}$$

and expand the trial function $\hat{u}_n$ as

$$\hat{u}_n(x,y) = \sum_{j=1}^{n} a_j \hat{\varphi}_j(x,y) + w(x,y) \tag{4.9.2}$$

in terms of the basis functions corresponding to the interior nodes. The sum of the first term on the right-hand side vanishes on the boundary ∂G, so that $\hat{u}_n(x,y)$ satisfies the boundary condition (4.7.2). An approximate solution $\hat{u}_n$ can be obtained by substituting (4.9.2) into the weak form (4.5.4).

4.10   An Example with a Mixed-Type Boundary Condition

In this section we consider a mixed-type boundary value problem with an inhomogeneous Dirichlet condition and a natural condition. Let the domain of the problem be G and assume that its boundary ∂G is divided into ∂G_1 and ∂G_2:

$$-\Delta u + qu = f \tag{4.10.1}$$

$$u = g \qquad \text{on } \partial G_1 \tag{4.10.2}$$

$$\frac{\partial u}{\partial n} + \alpha(x,y)u + \beta(x,y) = 0 \qquad \text{on } \partial G_2 \tag{4.10.3}$$

According to the procedure given in Sec. 4.7, we search for the finite element solution $\hat{u}_n$ in the form

$$\hat{u}_n(x,y) = \sum_{j=1}^{n} a_j \hat{\varphi}_j(x,y) + \sum_{j=n+1}^{n+\nu} g_j \hat{\varphi}_j(x,y) \tag{4.10.4}$$

The numbers from 1 to n correspond to the interior nodes, those from $n+1$ to $n+\nu$ correspond to the nodes on the boundary, and

$$g_j = g(x_j, y_j) \qquad j = n+1, \ldots, n+\nu \tag{4.10.5}$$

Substituting $\hat{u}_n$ for u in (4.10.1), multiplying by $\hat{\phi}_k$, $k = 1, 2, \ldots, n$, and integrating by parts using (4.10.3), we eventually obtain a system of linear equations with respect to $\{a_j\}$:

$$\sum_{j=1}^{n} \left(K_{kj} + M_{kj} + \int_{G_2} \alpha \hat{\varphi}_k \hat{\varphi}_j \, d\sigma \right) a_j = f_k - \sum_{j=n+1}^{n+\nu} (K_{kj} + M_{kj}) g_j$$

$$- \int_{\partial G_2} \beta \, \hat{\varphi}_k \, d\sigma \qquad k = 1, 2, \dots, n$$

$$(4.10.6)$$

The unknowns to be determined are n coefficients $a_j, j = 1, 2, \dots, n$, corresponding to the interior nodes and the nodes on the boundary ∂G_2. On the other hand, the conditions for determining these unknowns are the n linear equations (4.10.6) obtained by multiplication of $\hat{\phi}_k$ corresponding to these n nodes followed by integration. Thus the number of unknowns coincides with the number of conditions, so that we can solve these equations. Substitution of $\{a_j\}$ thus obtained into (4.10.4) results in a finite element solution.

We apply the procedure described above to the following simple model problem governed by Poisson's equation. Let domain G of the problem be the unit square $0 \leq x, y \leq 1$:

$$\frac{\partial^2 u}{\partial x^2} + \frac{\partial^2 u}{\partial y^2} = 1 \qquad (4.10.7)$$

$$u(x, 0) = 0 \qquad 0 \leq x \leq 1 \qquad (4.10.8)$$

$$u(x, 1) = x \qquad 0 \leq x \leq 1 \qquad (4.10.9)$$

$$\frac{\partial u}{\partial n}(0, y) = -\frac{\partial u}{\partial x}(0, y) = 0 \qquad 0 < y < 1 \qquad (4.10.10)$$

$$\frac{\partial u}{\partial n}(1, y) = \frac{\partial u}{\partial x}(1, y) = 0 \qquad 0 < y < 1 \qquad (4.10.11)$$

The boundary condition along the boundary ∂G_1 parallel to the x axis is of the Dirichlet type, and that along ∂G_2 parallel to the y axis is natural. This Dirichlet condition (4.10.9) is inhomogeneous, and α and β in (4.10.3) are zero.

We subdivide the domain G into triangular elements by dividing the intervals $(0, 1)$ of both x and y into eight equal subintervals and number the nodes as shown in Fig. 4.7. The nodes numbered $1, 2, \dots, 9$ and $73, 74, \dots, 81$ correspond to the boundary ∂G_1, while the nodes numbered $10, 19, \dots, 64$ and $18, 27, \dots, 63, 72$ correspond to the boundary ∂G_2. We write the finite element solution

$$u_n(x, y) = \sum_{j=10}^{72} a_j \varphi_j(x, y) + \left(\sum_{j=1}^{9} + \sum_{j=73}^{81} \right) g_j \hat{\varphi}_j(x, y) \qquad (4.10.12)$$

according to (4.10.4). The values of g_j at the nodes on ∂G_1 are given from

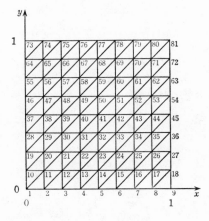

Fig. 4.7 Triangulation of the domain and the node numbers.

(4.10.8) and (4.10.9) as

$$g_j = 0 \qquad j = 1, 2, \ldots, 9$$
$$g_j = \tfrac{1}{8}(j - 73) \qquad j = 73, 74, \ldots, 81 \tag{4.10.13}$$

The unknowns to be determined are $n = 63$ coefficients $a_{10}, a_{11}, \ldots, a_{72}$ corresponding to the interior nodes and the boundary nodes on ∂G_2.

We obtain the finite element solution $\hat{u}_n$ by solving the system of linear equations (4.10.6) with respect to these unknowns. The solution $\hat{u}_n$ is shown in Fig. 4.8. It can be observed that the natural boundary condition $\partial u / \partial n = 0$ is approximately satisfied along the boundary ∂G_2.

4.11 *Methods for Solving a System of Linear Equations*

As we have seen so far, it is important in the FEM to use an efficient method for solving a system of linear equations. To present the details of methods for solving a system of linear equations is beyond the scope of

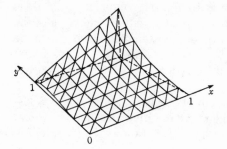

Fig. 4.8 A finite element solution $\hat{u}_n(x, y)$.

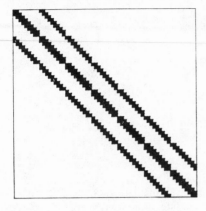

Fig. 4.9 Distribution of the possible nonzero entries of the coefficient matrix.

this book, and so we leave it to other books and restrict ourselves only to some short comments on these procedures.

One of the most remarkable characteristics of the coefficient matrix of the system of linear equations encountered in applications of the FEM is that it is very large and sparse; that is, the ratio of nonzero entries is relatively small. For example, in the simple model problem given in the previous section, the coefficient matrix is 63×63 and the total number of entries is 3969. It is obvious that if we repeat the mesh refinement the dimensions of the matrix will become explosively large. However, the ratio of nonzero entries is small, and they are concentrated in the neighborhood of the diagonal. In Fig. 4.9 the distribution of possible nonzero entries of the matrix in our example due to the adjacency of the nodes is shown. The small black squares represent possible nonzero entries. Their total number is 415, which is only about 10.5 percent of the entire number of entries.

Thus the coefficient matrix of the system of linear equations encountered in the FEM has a banded form, and such a matrix is called a *band matrix*. The distance from the diagonal to the most distant nonzero entry is called the *half-bandwidth*. If w is the half-bandwidth, then $2w + 1$ is the *bandwidth* of the matrix if it is symmetric. In the example given in the previous section, the half-bandwidth is 10 and the bandwidth is 21. It is evident that the half-bandwidth is closely related to the numbering of the nodes. The half-bandwidth of the matrix in the example is 10 because the number of the nodes has a period of 9. From this example we can see that the half-bandwidth becomes smaller if we assign numbers closer to the adjacent nodes.

The methods used to solve a system of linear equations can be classified roughly into the direct method and the iterative method. The *direct method* is a procedure in which the arithmetic operations are executed directly on the entries of the coefficient matrix and the solution is obtained in finite steps. The most important and most common of the direct methods is the *Gauss elimination method*. This procedure is often applied in such a way that the coefficient matrix is decomposed into a lower left matrix *L* and an upper right matrix *U* and is sometimes called *LU decomposition*. Usually a subroutine of the Gauss elimination method is coded for a general matrix regardless of the distribution of nonzero entries. The matrix arising in the FEM, on the other hand, is generally a large, sparse band matrix, hence in the FEM a subroutine suitably modified for such a band matrix should be used in order to improve computational efficiency and save computer memory.

The *iterative method* for solving a system of linear equations is a procedure in which an iterative matrix is constructed by decomposing the given matrix into an appropriate form. It is applied repeatedly to an initial vector approaching the solution. Among such procedures the *Gauss-Seidel method* and the *successive over relaxation* (SOR) *method* are well known. It is easy to write an efficient program for an iterative method by taking into account the sparseness of the coefficient matrix. However, with the SOR method, it is usually difficult to determine the relaxation parameter theoretically in the FEM, and in order to obtain an appropriate parameter, for example, numerical experiments must be carried out beforehand.

A method called the *conjugate gradient* (CG) *method* may be regarded as one of the direct methods in principle and as one of the iterative methods in practice. Although the original CG method itself is not very efficient, appropriate preconditioning of the original matrix followed by the CG method results in remarkable improvement. For example, an incomplete LU decomposition in which the zero entries of the original matrix are forced to remain zero during the LU decomposition is one of the efficient preconditioning steps for the CG method.

The form of the coefficient matrix encountered in practical problems changes significantly depending on the type of differential equation, the shape of the domain, and the boundary condition. Although the Gauss elimination method and its modifications seem to be used fairly often, knowledge and experience are necessary for the user to choose the best method for a given problem.

5

Computation of Matrix Elements and Coordinate Transformation

5.1 *Element Matrices*

When we formulate the FEM or investigate it mathematically, the basis function plays an important role. However, when we apply the FEM to problems in practice, it is more convenient to handle a segment of a basis function restricted within a triangular element than to handle a basis function as a whole. In other words, in computing the entries of the coefficient matrix and the vector on the right-hand side of a particular equation, it is more convenient from a practical point of view to determine the contributions to each entry from each triangular element and assemble them into the global matrix.

Consider a triangular element and denote it by τ. Let the vertices or the nodes of this triangle be i, j, and k as shown in Fig. 5.1. Then, only the basis functions $\hat{\phi}_i$, $\hat{\phi}_j$, and $\hat{\phi}_k$ do not vanish in τ. On the other hand, the finite element solution $\hat{u}_n$ restricted within τ based on piecewise linear polynomials can be written as

$$\hat{u}_n(x, y)\bigg|_\tau = a_i \hat{\varphi}_i + a_j \hat{\varphi}_j + a_k \hat{\varphi}_k \qquad (5.1.1)$$

Therefore the contributions to τ from the integrals of the derivative term and the term including q are

$$\iint_\tau \left(\frac{\partial \hat{u}_n}{\partial x} \frac{\partial \hat{\varphi}_l}{\partial x} + \frac{\partial \hat{u}_n}{\partial y} \frac{\partial \hat{\varphi}_l}{\partial y} \right) dx\, dy$$

$$= a_i k_{il}^\tau + a_j k_{jl}^\tau + a_k k_{kl}^\tau \qquad l = i, j, k \qquad (5.1.2)$$

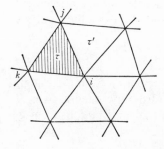

Fig. 5.1 The triangular element τ and nodes.

and

$$\iint_{\tau} q\hat{u}_n\hat{\varphi}_l\,dx\,dy = a_i m_{il}^{\tau} + a_j m_{jl}^{\tau} + a_k m_{kl}^{\tau} \qquad l = i,j,k \qquad (5.1.3)$$

respectively, where $k_{\mu\nu}^{\tau}$ and $m_{\mu\nu}^{\tau}$ are the $\mu\nu$ entries of the essentially 3×3 matrices k^{τ} and m^{τ}, respectively, defined by

$$k_{\mu\nu}^{\tau} = \iint_{\tau} \left(\frac{\partial\hat{\varphi}_{\mu}}{\partial x}\frac{\partial\hat{\varphi}_{\nu}}{\partial x} + \frac{\partial\hat{\varphi}_{\mu}}{\partial y}\frac{\partial\hat{\varphi}_{\nu}}{\partial y} \right) dx\,dy \qquad (5.1.4)$$

$$m_{\mu\nu}^{\tau} = \iint_{\tau} q\hat{\varphi}_{\mu}\hat{\varphi}_{\nu}\,dx\,dy \qquad (5.1.5)$$

Note that, once the triangulation of the domain G is fixed, these entries need be computed only once for each triangular element since they are independent of the particular boundary condition. This kind of matrix is called an *element matrix*. More specifically, k^{τ} and m^{τ} are called the *element stiffness matrix* and the *element mass matrix*.

Similarly, the contribution to τ of the integral including f is given by

$$\iint_{\tau} f\varphi_l\,dx\,dy = f_l^{\tau} \qquad l = i,j,k \qquad (5.1.6)$$

where f_l^{τ} is the lth entry of the vector $\mathbf{f}^{\tau}$ that consists of three entries. Contributions to τ from the integrals appearing in (4.5.4) are mentioned without omission as above.

5.2 Assembling Matrices

The next step is to assemble the element matrices k^{τ} and m^{τ} and the vectors $\mathbf{f}^{\tau}$ into the global matrices and the vector of (4.5.8) over the entire domain. To this end we focus on the number (i,j) of nodes instead of the number τ of triangular elements. That is, the ij entries of the stiffness matrix K or the mass matrix M as a whole can be obtained by summing up the contributions from the triangular elements adjacent to both nodes i and j. When piecewise linear basis functions are employed, the number of such triangular elements is only 2. For example, the triangular elements adjacent to both nodes i and j in Fig. 5.1 are τ and τ', so that the ij entry of K is given by

$$K_{ij} = k^\tau_{ij} + k^{\tau'}_{ij} \tag{5.2.1}$$

The first term on the right-hand side is the contribution from the integral over τ, and the second term is the contribution from the integral over τ'. Similarly, in Fig. 5.1, K_{ii} is given by the sum of the contributions from the six triangular elements adjacent to node i:

$$K_{ii} = \sum_\tau k^\tau_{ii} \tag{5.2.2}$$

We summarize here the procedure mentioned above, including the process for assembling the right-hand side vector $\mathbf{f}$:

$$K_{ij} = \sum_\tau k^\tau_{ij} \tag{5.2.3}$$

$$M_{ij} = \sum_\tau m^\tau_{ij} \tag{5.2.4}$$

$$f_j = \sum_\tau f^\tau_j \tag{5.2.5}$$

These relations can also be verified directly by decomposing (4.5.10), (4.5.11), and (4.5.9), respectively, into the contributions from each triangular element.

In actual computation, we first divide the entire domain G into small triangular elements and compute k^τ_{ij}, m^τ_{ij}, and f^τ_j for each element τ. Then, if the three nodes of τ are i, j, and k, we add $k^\tau_{\mu\nu}$ to $K_{\mu\nu}$, $m^\tau_{\mu\nu}$ to $M_{\mu\nu}$, and f^τ_μ to f^τ, where (μ, ν) is taken over all $3 \times 3 = 9$ pairs including (i,i), (j,j), and (k,k).

5.3 Linear Shape Functions

As already mentioned, from a practical point of view it is more convenient to use the segment of the basis function restricted within each triangular element than to use the basis function $\hat{\phi}_j(x,y)$ as a whole when computing the components of the matrices and the vector. In the case of the piecewise linear basis function, the components $k^\tau_{\mu\nu}$, $m^\tau_{\mu\nu}$, and f^τ_μ of each triangular element τ can be computed if the segment $\xi_i(x,y)$ of the basis function $\hat{\phi}_i(x,y)$ restricted within τ (Fig. 5.2A), the segment $\xi_j(x,y)$ of $\hat{\phi}_j$ within τ (Fig. 5.2B), and the segment $\xi_k(x,y)$ of $\hat{\phi}_k$ within τ (Fig. 5.2C) are known. Each of the three functions ξ_i, ξ_j, and ξ_k shown in Fig. 5.2 is

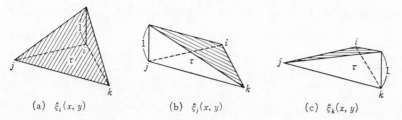

(a) $\xi_i(x, y)$ (b) $\xi_j(x, y)$ (c) $\xi_k(x, y)$

Fig. 5.2 Shape functions of first order.

called a *shape function* in τ. They are shape functions of first order. They are also called *coordinate functions* because $\hat{u}_n$ can be expressed as

$$\hat{u}_n = a_i\xi_i + a_j\xi_j + a_k\xi_k \qquad (5.3.1)$$

in τ.

We give here the specific forms of the linear shape functions. Hereafter we denote the three vertices of the triangular elements τ by P_1, P_2, and P_3, and the coordinates of these vertices by (x_1, y_1), (x_2, y_2), and (x_3, y_3), respectively. Then the linear shape function in τ can be written as

$$\xi_i(x, y) = \frac{1}{2S} \begin{vmatrix} 1 & 1 & 1 \\ x & x_j & x_k \\ y & y_j & y_k \end{vmatrix}$$

$$= \frac{1}{2S}\{(x_jy_k - x_ky_j) + (y_j - y_k)x - (x_j - x_k)y\} \qquad (5.3.2)$$

where

$$S = \frac{1}{2}\begin{vmatrix} 1 & 1 & 1 \\ x_1 & x_2 & x_3 \\ y_1 & y_2 & y_3 \end{vmatrix} \qquad (5.3.3)$$

The index (i, j, k) runs over the three permutations $(1, 2, 3), (2, 3, 1)$, and $(3, 1, 2)$. As is well known, the absolute value of S is equal to the area of the triangle $P_1P_2P_3$. If P_1, P_2, and P_3 are located counterclockwise, S becomes positive. It is evident that $\xi_i(x, y)$ satisfies the property stated above; that is, it is a linear function satisfying

$$\xi_i(x_l, y_l) = \begin{cases} 1 & l = i \\ 0 & l \neq i \end{cases} \qquad (5.3.4)$$

The derivatives of $\xi_i(x,y)$ are given from (5.3.2) as

$$\frac{\partial \xi_i}{\partial x} = \frac{1}{2S} (y_j - y_k) \qquad (5.3.5)$$

$$\frac{\partial \xi_i}{\partial y} = -\frac{1}{2S} (x_j - x_k) \qquad (5.3.6)$$

Substitution of the specific form (5.3.2) of the shape function into (5.3.1) leads to an expression of $\hat{u}_n$ in the triangular element τ:

$$\hat{u}_n = \left\{ \begin{vmatrix} a_i & a_j & a_k \\ x_i & x_j & x_k \\ y_i & y_j & y_k \end{vmatrix} - \begin{vmatrix} a_i & a_j & a_k \\ 1 & 1 & 1 \\ y_i & y_j & y_k \end{vmatrix} x \right.$$

$$\left. - \begin{vmatrix} a_i & a_j & a_k \\ x_i & x_j & x_k \\ 1 & 1 & 1 \end{vmatrix} y \right\} \Big/ \begin{vmatrix} 1 & 1 & 1 \\ x_i & x_j & x_k \\ y_i & y_j & y_k \end{vmatrix} \qquad (5.3.7)$$

5.4 Quadratic Shape Functions

When solving problems including high-order derivatives, shape functions of order more than 1 may be used, as will be mentioned later. In this section we present quadratic shape functions that take the specified values at the three vertices P_1, P_2, and P_3 of the triangular element τ and the midpoints P_4, P_5, and P_6 of the corresponding opposite sides, as shown Fig. 5.3.

First, the quadratic shape function $\xi_1^{(2)}$, which takes the value unity at node P_1 and the value zero at the other five nodes, is expressed as

$$\xi_1^{(2)}(x,y) = \xi_1(2\xi_1 - 1) \qquad (5.4.1)$$

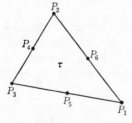

Fig. 5.3 The triangular element τ and nodes for shape functions of second order.

Since ξ_1 is a linear function that takes the value unity at P_1 and the value zero at P_2, it is obvious that ξ_1 takes the value $\frac{1}{2}$ at P_6. Similarly, we can construct $\xi_2^{(2)}$ and $\xi_3^{(2)}$, which take the value unity at nodes P_2 and P_3, respectively. Next, the quadratic shape function $\xi_4^{(2)}$, which takes the value unity at node P_4 and the value zero at the other five nodes, is given by

$$\xi_4^{(2)}(x, y) = 4\xi_2\xi_3 \tag{5.4.2}$$

Both $\xi_5^{(2)}$ and $\xi_6^{(2)}$ can be obtained in the same way. It is easy to see that the shape functions given above are quadratic functions such that

$$\xi_j^{(2)}(x_k, y_k) = \begin{cases} 1 & j = k \\ 0 & j \neq k \end{cases} \tag{5.4.3}$$

For the finite element solution, a quadratic trial function in the triangular element τ can be expressed in terms of these shape functions as

$$\hat{u}_n(x, y) = \sum_{j=1}^{6} a_j\xi_j^{(2)}(x, y) \tag{5.4.4}$$

This expression satisfies

$$\hat{u}_n(x_j, y_j) = a_j \tag{5.4.5}$$

as does (4.5.2).

5.5 Transformation Into a Standard Triangle

When we compute the components of the element matrices or the vectors in a triangular element, a coordinate transformation into a standard triangle with local coordinates often makes the computation quite efficient. In particular, when $f(x, y)$ is a complicated function of x and y such that it is impossible to integrate it analytically, we must inevitably employ a numerical integration formula defined by local coordinates, hence a transformation is indispensable. In such cases, we first transform the given triangular element in global xy coordinates into a standard triangle in local coordinates, compute the integral approximately using a numerical integration formula defined in the local coordinates, and then transform the result back into the original global xy coordinates.

Here we choose the right triangle $Q_1Q_2Q_3$ (Fig. 5.4B) whose vertices are $Q_1(1,0)$, $Q_2(0,1)$, and $Q_3(0,0)$ in the local $\xi_1\xi_2$ coordinate system as the *standard triangle* T into which the given triangular element $P_1P_2P_3$ in the global xy-coordinate system (Fig. 5.4A) is transformed. In order to write the coordinate transformation explicitly we make use of the shape

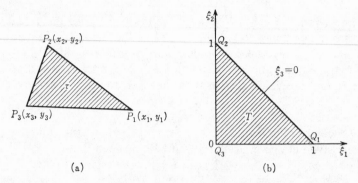

Fig. 5.4 The triangular element τ and the standard triangle T.

functions (5.3.2). That is, the transformation is given by

$$\xi_1 = \frac{1}{2S} \begin{vmatrix} 1 & 1 & 1 \\ x & x_2 & x_3 \\ y & y_2 & y_3 \end{vmatrix} = \frac{1}{2S} \{(x_2 y_3 - x_3 y_2) + (y_2 - y_3)x - (x_2 - x_3)y\}$$

$$(5.5.1)$$

$$\xi_2 = \frac{1}{2S} \begin{vmatrix} 1 & 1 & 1 \\ x & x_3 & x_1 \\ y & y_3 & y_1 \end{vmatrix} = \frac{1}{2S} \{(x_3 y_1 - x_1 y_3) + (y_3 - y_1)x - (x_3 - x_1)y\}$$

where S is defined by (5.3.3). It is evident from the relation (5.3.4) that the transformation (5.5.1) from the global coordinate system (x, y) into the local coordinate system (ξ_1, ξ_2) is a linear transformation that maps the points P_1, P_2, and P_3 onto the points Q_1, Q_2, and Q_3, respectively.

Since ξ_1, ξ_2, and ξ_3 are linear functions satisfying (5.3.4), their sum is identical unity, that is,

$$\xi_1 + \xi_2 + \xi_3 = 1 \qquad (5.5.2)$$

Hence, if we take into account ξ_3, which does not appear explicitly in the transformation, we see that the side $Q_1 Q_2$ is given by $\xi_3 = 0$ from (5.5.2). It is evident that the side $Q_3 Q_1$ is given by $\xi_2 = 0$ and that the side $Q_2 Q_3$ is given by $\xi_1 = 0$.

The inverse transformation of (5.5.1) is given by

$$x = x(\xi_1, \xi_2) = (x_1 - x_3)\xi_1 + (x_2 - x_3)\xi_2 + x_3$$
$$y = y(\xi_1, \xi_2) = (y_1 - y_3)\xi_1 + (y_2 - y_3)\xi_2 + y_3$$
$$(5.5.3)$$

which can be verified through point-to-point correspondence. Therefore the Jacobian J of the present transformation is

$$J = \begin{vmatrix} x_1 - x_3 & x_2 - x_3 \\ y_1 - y_3 & y_2 - y_3 \end{vmatrix} = 2S \tag{5.5.4}$$

A definite integral of a function $g(x, y)$ over the triangular element τ is transformed by (5.5.3):

$$\iint_\tau g(x, y) \, dx \, dy = \iint_T g(x(\xi_1, \xi_2), y(\xi_1, \xi_2)) J \, d\xi_1 \, d\xi_2 \tag{5.5.5}$$

Note that the Jacobian J in the present transformation is a constant.

5.6 Computation of Integrals in the Barycentric Coordinate System

The coordinate system over a local triangular domain, introduced in the preceding section, is called the *barycentric coordinate system* or the *area coordinate system*. In this coordinate system the position of a point P in the triangular element τ is specified by the coordinate (ξ_1, ξ_2, ξ_3) consisting of three components. It is called the area coordinate system because the following relation holds for the triangle shown in Fig. 5.5:

$$\xi_1 = \frac{\text{length of } PP'}{\text{length of } P_iP_i'} = \frac{\text{area of } \Delta PP_jP_k}{\text{area of } \Delta P_1P_2P_3} \tag{5.6.1}$$

Needless to say, only two of the ξ_1, ξ_2, and ξ_3 components are independent. The relation between the coordinate system (x, y) and the coordinate system (ξ_1, ξ_2, ξ_3) is given by (5.5.2) and (5.5.3), and the latter can be expressed in the matrix form as

$$\begin{pmatrix} 1 \\ x \\ y \end{pmatrix} = \begin{pmatrix} 1 & 1 & 1 \\ x_1 & x_2 & x_3 \\ y_1 & y_2 & y_3 \end{pmatrix} \begin{pmatrix} \xi_1 \\ \xi_2 \\ \xi_3 \end{pmatrix} \tag{5.6.2}$$

If we solve this equation inversely for ξ_1, ξ_2, and ξ_3, we obtain (5.3.2).

If the domain of the given integral is a triangle and the integrand is a

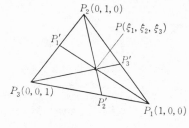

$P_2(0, 1, 0)$
$P(\xi_1, \xi_2, \xi_3)$
P_1'
P_3'
$P_3(0, 0, 1)$
P_2'
$P_1(1, 0, 0)$

Fig. 5.5 The area coordinates (ξ_1, ξ_2, ξ_3).

polynomial or a power function of ξ_1, ξ_2, and ξ_3, the following formula derived from (5.5.5) will be useful:

$$\iint_\tau \xi_1^\alpha \xi_2^\beta \xi_3^\gamma \, dx \, dy = \iint_T \xi_1^\alpha \xi_2^\beta \xi_3^\gamma I \, d\xi_1 \, d\xi_2$$

$$= \frac{\Gamma(\alpha+1)\Gamma(\beta+1)\Gamma(\gamma+1)}{\Gamma(\alpha+\beta+\gamma+3)} 2S \tag{5.6.3}$$

where $\alpha, \beta, \gamma > -1$ and $\Gamma(\mu+1)$ is the gamma function defined by

$$\Gamma(\mu+1) = \int_0^\infty t^\mu e^{-t} \, dt \tag{5.6.4}$$

which is $\Gamma(\mu+1) = \mu!$ if μ is equal to zero or a positive integer. The formula (5.6.3) is easily verified by substituting $\xi_3 = 1 - \xi_1 - \xi_2$ on the right-hand side and integrating over T using the relation

$$\int_0^1 \zeta^\mu (1-\zeta)^\nu \, d\zeta = \frac{\Gamma(\mu+1)\,\Gamma(\nu+1)}{\Gamma(\mu+\nu+2)} \tag{5.6.5}$$

5.7 Entries of Element Matrices and Triangulation of the Acute Type

From (5.6.3) the element mass matrix m^τ of (5.1.5) in the case of $q = 1$ is

$$m_{ij}^\tau = \iint_\tau \hat\varphi_i \hat\varphi_j \, dx \, dy = \iint_\tau \xi_i \xi_j \, dx \, dy = 2S \iint_T \xi_i \xi_j \, d\xi_1 \, d\xi_2$$

$$= \begin{cases} \dfrac{1}{6} S & i = j \\[2mm] \dfrac{1}{12} S & i \ne j \end{cases} \tag{5.7.1}$$

The element stiffness matrix k^τ can be computed from the derivative of (5.5.1) as

$$k_{ij}^\tau = \iint_\tau \left(\frac{\partial \xi_i}{\partial x} \frac{\partial \xi_j}{\partial x} + \frac{\partial \xi_i}{\partial y} \frac{\partial \xi_j}{\partial y} \right) dx \, dy$$

$$= \frac{1}{4S} \{ (x_j - x_k)(x_k - x_i) + (y_j - y_k)(y_k - y_i) \} \qquad i \ne j$$

$$= \frac{1}{4S} q_i^T q_j = \begin{cases} \dfrac{1}{4S} |q_i|^2 & i = j \\[2mm] -\dfrac{1}{4S} |q_i| \, |q_j| \cos \theta_k & i \ne j \end{cases} \tag{5.7.2}$$

where q_i, q_j, and q_k are the vectors of the sides $P_j P_k$, $P_k P_i$, and $P_i P_j$, re-

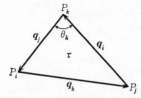

Fig. 5.6 The triangular element τ and the vectors of the sides.

spectively, of the triangular element $P_i P_j P_k$ and θ_k is the interior angle at the vertex P_k.

From this relation we see that the off-diagonal entries of the element stiffness matrix are zero or negative unless the triangle τ has an obtuse angle. If all the triangular elements have no obtuse angle, then the triangulation is said to be of the *acute type*. Acute-type triangulation plays an important role in the stability of the FEM when it is applied to some time-dependent problems.

5.8 *Numerical Integration*

If the given integral is not simple enough to be computed analytically, an approximate value may be obtained by applying to the right-hand side of (5.5.5) an appropriate numerical integration formula defined on the standard triangle T:

$$\frac{\iint_T G(\xi_1, \xi_2)\, d\xi_1\, d\xi_2}{\sum_{j=1}^{M} B_j\, G(\alpha_j, \beta_j)} \tag{5.8.1}$$

where (α_j, β_j), $j = 1, 2, \ldots, M$, are the points and B_j are the corresponding weights of the formula.

By transforming this formula back to the original triangular element τ using

$$G(\xi_1, \xi_2) = g(x, y)J \tag{5.8.2}$$

we obtain a numerical integration formula over τ:

$$\frac{\iint_\tau g(x, y)\, dx\, dy}{\sum_{j=1}^{M} A_j g(a_j, b_j)} \tag{5.8.3}$$

$$a_j = x(\alpha_j, \beta_j) \qquad b_j = y(\alpha_j, \beta_j) \tag{5.8.4}$$

$$A_j = B_j J \tag{5.8.5}$$

A simple example of a numerical integration formula is given in Sec. 6.7. In practical problems Gauss-type formulas are widely used.

5.9 Isoparametric Transformation

So far we have assumed that the shape of the given domain G is a polygon. However, the boundaries encountered in practical problems are usually curved. In this section we will discuss the isoparametric transformation, which is useful when dealing with a curved boundary.

We have introduced a coordinate transformation from an arbitrary triangle into a standard triangle using a linear mapping function. However, if a side of the given triangle is curved, it is impossible to transform it into the standard triangle using a linear function. In order to transform a curved side into a line segment, a mapping function that is at least quadratic is necessary. In this section we assume that the given boundary is piecewise quadratic, so that the image of a quadratic curved segment produced by a quadratic transformation becomes precisely a line segment. Then we consider the transformation of a triangle $P_1 P_2 P_3$, given in global xy coordinates, whose sides consist of quadratic curves, as shown in Fig. 5.7A, into the standard triangle $Q_1 Q_2 Q_3$ in local $\xi_1 \xi_2$ coordinates as shown in Fig. 5.7B.

As already seen in (5.4.5), a quadratic function that takes the prescribed values $u_j, j = 1, 2, \ldots, 6$, at the three vertices Q_1, Q_2, and Q_3 and the midpoints Q_4, Q_5, and Q_6 of the three sides is given by

$$u = \sum_{j=1}^{6} u_j \xi_j^{(2)}(\xi_1, \xi_2) \tag{5.9.1}$$

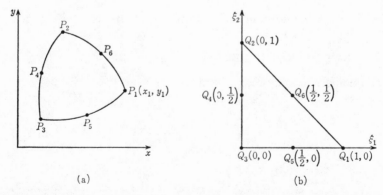

Fig. 5.7 The triangular element $P_1 P_2 P_3$ with curved sides and the standard triangle $Q_1 Q_2 Q_3$.

where $\xi_j^{(2)}$, $j = 1, 2, \ldots, 6$, are the quadratic shape functions given by (5.4.1), (5.4.2), and so on.

Now from the form of (5.9.1) we see that we can divert it as a quadratic transformation giving the correspondence between the point in xy coordinates and the point in $\xi_1 \xi_2$ coordinates. Thus we can define a quadratic transformation between xy coordinates and $\xi_1 \xi_2$ coordinates in exactly the same form as (5.9.1) by

$$x = \sum_{j=1}^{6} x_j \xi_j^{(2)}(\xi_1, \xi_2)$$

$$y = \sum_{j=1}^{6} y_j \xi_j^{(2)}(\xi_1, \xi_2)$$

$$(5.9.2)$$

where $(x_1, y_1), \ldots, (x_6, y_6)$ are the coordinates of the three vertices and the midpoints of the three curved sides of the triangle in xy coordinates. It is evident that point Q_j in Fig. 5.7B corresponds to point P_j in Fig. 5.7A. This transformation is called an *isoparametric transformation*. The basic idea of an isoparametric transformation is that the point (x, y) in xy coordinates is transformed by (5.9.2) to the point (ξ_1, ξ_2) in $\xi_1 \xi_2$ coordinates, while the function values are also transformed by (5.9.1), which has the same form as the point transformation (5.9.2).

5.10 Transformation of a Triangular Element With One Curved Side

In the actual application of the FEM triangular elements with a curved side appear along the boundary of the domain, and each of them usually has two straight sides in the interior of the domain and one curved side along the boundary. We also assume that every curved side consists of a quadratic curve. Then it is convenient to define the $\alpha\beta$ coordinates intermediately and to consider a transformation of the triangle with a curved side shown in Fig. 5.8A into the triangle with $\alpha\beta$ coordinates shown in Fig. 5.8B. For this transformation we take the linear function (5.5.1), which maps P_1, P_2, and P_3 onto R_1, R_2, and R_3, respectively. We denote the image points $R_4(0, \frac{1}{2})$, $R_5(\frac{1}{2}, 0)$, and $R_6(\alpha_6, \beta_6)$ of this transformation by $P_4(x_4, y_4)$, $P_5(x_5, y_5)$, and $P_6(x_6, y_6)$. The Jacobian of this transformation is given by (5.5.4), and nothing new is encountered here.

Next we consider a quadratic transformation that maps the triangle $R_1 R_2 R_3$ in $\alpha\beta$ coordinates (Fig. 5.8B) onto the standard triangle $Q_1 Q_2 Q_3$ in $\xi_1 \xi_2$ coordinates (Fig. 5.8C). This transformation is given by

$$\alpha = 2(2\alpha_6 - 1)\xi_1\xi_2 + \xi_1$$

$$\beta = 2(2\beta_6 - 1)\xi_1\xi_2 + \xi_2$$

$$(5.10.1)$$

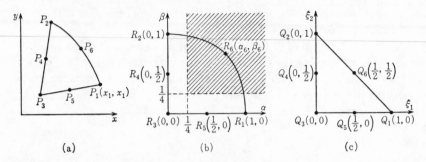

Fig. 5.8 Transformation of a triangle with one curved side.

It is easy to see that by this transformation the points $Q_j, j = 1, 2, \ldots, 6$, are mapped onto the points $R_j, j = 1, 2, \ldots, 6$, respectively. This is a particular case of (5.9.2). The Jacobian J of this transformation is given by

$$
\begin{aligned}
J &= J(\xi_1, \xi_2) \\
&= 1 + 2(2\beta_6 - 1)\xi_1 + 2(2\alpha_6 - 1)\xi_2
\end{aligned}
\tag{5.10.2}
$$

If the Jacobian happens to vanish at any point in the $\xi_1 \xi_2$ plane, this transformation cannot be used. If we assume, on the other hand, $J(1,0) > 0$ and $J(0,1) > 0$, then, since $J(0,0) = 1 > 0$, J becomes positive inside the triangle $Q_1 Q_2 Q_3$ and the transformation can be used. This assumption is equivalent to

$$
\alpha_6 > \tfrac{1}{4} \qquad \beta_6 > \tfrac{1}{4}
\tag{5.10.3}
$$

That is, if the point R_6 is located in the domain with oblique lines in Fig. 5.8B, or, in other words, if P_6 is located near the midpoint of the curved side, the Jacobian of this transformation never vanishes.

An inverse transformation of (5.10.1) is obtained when we solve

$$
2(2\beta_6 - 1)\xi_1^2 + \{2(2\alpha_6 - 1)\beta - 2(2\beta_6 - 1)\alpha + 1\}\xi_1 - \alpha = 0
$$

$$
2(2\alpha_6 - 1)\xi_2^2 + \{2(2\beta_6 - 1)\alpha - 2(2\alpha_6 - 1)\beta + 1\}\xi_2 - \beta = 0
$$

$$
\tag{5.10.4}
$$

for ξ_1 and ξ_2. Thus, if an integral of the element matrix given in xy coordinates is transformed into an integral in $\xi_1 \xi_2$ coordinates, it usually becomes complicated and a numerical integration formula must inevitably be employed.

6

Error Analysis of the Finite Element Method in Two Space Dimensions and Variational Crimes

6.1 *Interpolation Over a Triangle*

The purpose of this chapter is to present an error analysis of the FEM in two space dimensions. In the error analysis in one space dimension we introduced interpolation as an intermediate tool. In this section we also first introduce interpolation over a triangle.

Suppose that the function $u(x,y)$ is sufficiently smooth and takes the values

$$u(x_l, y_l) = u_l \qquad l = 1,2,3 \tag{6.1.1}$$

at the vertices $P_1(x_1,y_1)$, $P_2(x_2,y_2)$, and $P_3(x_3,y_3)$ of the triangular element τ. Although in the function u we are actually seeking the exact solution of the given problem, for the moment we regard $u(x,y)$ not as the particular function but as a general function. Corresponding to the function u we define $\hat{u}_1(x,y)$ expressed in terms of a linear combination of the shape functions

$$\hat{u}_1(x,y) = u_1\xi_1(x,y) + u_2\xi_2(x,y) + u_3\xi_3(x,y) \tag{6.1.2}$$

satisfying

$$\hat{u}_1(x_l, y_l) = u_l \qquad l = 1,2,3 \tag{6.1.3}$$

which means that $\hat{u}_1(x,y)$ is a linear interpolate that equals $u(x,y)$ at the three points P_1, P_2, and P_3.

71

6.2 *Error Estimation of the Interpolation*

This section describes the error

$$\epsilon_1(x,y) = u(x,y) - \hat{u}_1(x,y) \tag{6.2.1}$$

of the interpolate $\hat{u}_1(x,y)$. Although we discussed mean square error in Sec. 3.8, we consider here pointwise maximum error for simplicity. Let $P_0(x_0,y_0)$ be a point inside the triangular element τ and let

$$h_1 = x - x_0 \qquad h_2 = y - y_0 \tag{6.2.2}$$

Then the Taylor expansion of u around P_0 becomes

$$u(x,y) = p_1(x,y) + R_2(x,y) \tag{6.2.3}$$

where

$$p_1(x,y) = u(x_0,y_0) + h_1\frac{\partial u}{\partial x}(x_0,y_0) + h_2\frac{\partial u}{\partial y}(x_0,y_0) \tag{6.2.4}$$

$$R_2(x,y) = \frac{1}{2!}\left(h_1\frac{\partial}{\partial x} + h_2\frac{\partial}{\partial y}\right)^2 u(x_0+\theta h_1, y_0+\theta h_2) \qquad 0<\theta<1 \tag{6.2.5}$$

The differential operator appearing in the residual R_2 is defined recursively by

$$\left(h_1\frac{\partial}{\partial x} + h_2\frac{\partial}{\partial y}\right)^k u = \left(h_1\frac{\partial}{\partial x} + h_2\frac{\partial}{\partial y}\right)\left(h_1\frac{\partial}{\partial x} + h_2\frac{\partial}{\partial y}\right)^{k-1} u \tag{6.2.6}$$

and $P_1(x,y)$ is obviously a linear polynomial. Let the maximum length of the sides of the triangular element τ be h. Then, since $|h_1|,|h_2| \le h$, we have an estimate from (6.2.5):

$$|R_2(x,y)| \le C_1' h^2 \max_{\substack{(x,y)\in\tau \\ |m|=2}} |D^m u| \tag{6.2.7}$$

where C_1' is a constant independent of h. We also assume that we can take C' such that it does not depend on the choice of the triangular element τ. In (6.2.7) D is the differential operator defined in multiindex notation such that, for a pair $m = (\mu,\nu)$,

$$D^m u = \frac{\partial^{|m|}u}{\partial x^\mu \partial y^\nu} \tag{6.2.8}$$

where $|m| = \mu + \nu$. This is a conventional expression for several derivatives defined on the right-hand side in a single form. The inequality (6.2.7) indicates, in short, that if the second derivative of u is bounded, then the error of the interpolate is of order h^2.

We replace u by its derivative Du, that is, by $\partial u/\partial x$ or $\partial u/\partial y$. Then we have the Taylor expansion of the derivative of u.

$$Du(x,y) = p_0 + R_1(x,y) \qquad (6.2.9)$$

where

$$p_0 = Du(x_0, y_0) = Dp_1(x_0, y_0) \qquad (6.2.10)$$

$$R_1(x,y) = \left(h_1 \frac{\partial}{\partial x} + h_2 \frac{\partial}{\partial y}\right) Du(x_0 + \theta' h_1, y_0 + \theta' h_2) \qquad 0 < \theta' < 1$$
$$(6.2.11)$$

Note that p_0 is a constant in the present case. From (6.2.11) we have an estimate for the residual R_1:

$$|R_1(x,y)| \leq C_0' h \max_{\substack{(x,y) \in \tau \\ |m|=2}} |D^m u| \qquad (6.2.12)$$

where we also assume that C_0' does not depend on τ.

The interpolate $\hat{u}_1(x,y)$ is a linear function inside the triangular element τ which can be decomposed as

$$\hat{u}_1(x,y) = p_1(x,y) + R_1(x,y) \qquad (6.2.13)$$

where $p_1(x,y)$ is the linear function defined by (6.2.4). Since $\hat{u}_1$ and p_1 are linear, the residual R_1 is also a linear function in the present case. Moreover, u equals $\hat{u}_1$ at the vertices $P_1(x_1, y_1)$, $P_2(x_2, y_2)$, and $P_3(x_3, y_3)$, so that R_2 also agrees with R_1 there. Therefore the linear function R_1 can be written as

$$R_1(x,y) = R_2(x_1, y_1)\xi_1(x,y) + R_2(x_2, y_2)\xi_2(x,y) + R_2(x_3, y_3)\xi_3(x,y)$$
$$(6.2.14)$$

Since by definition the shape functions ξ_i, $i = 1, 2, 3$, satisfy

$$|\xi_i(x,y)| \leq 1 \qquad i = 1, 2, 3 \qquad (6.2.15)$$

we have, from (6.2.7) and (6.2.14),

$$|R_1(x,y)| \leq 3C_1' h^2 \max_{\substack{(x,y) \in \tau \\ |m|=2}} |D^m u| \qquad (6.2.16)$$

On the other hand, from (6.2.3) and (6.2.13) we have

$$\epsilon_1(x,y) = u - \hat{u}_1 = R_2 - R_1 \qquad (6.2.17)$$

Thus we eventually obtain the following error estimate for the interpolation:

$$|\epsilon_1(x,y)| \leq |R_2(x,y)| + |R_1(x,y)|$$
$$\leq C_1 h^2 \max_{\substack{(x,y) \in \tau \\ |m|=2}} |D^m u| \qquad (6.2.18)$$

6.3 *Error Estimation of the Derivative of the Interpolate*

The derivative of the interpolate becomes, from (6.2.13) and (6.2.10),

$$D\hat{u}_1(x, y) = p_0 + DR_1(x, y) \qquad (6.3.1)$$

The error term on the right-hand side is estimated from (6.2.14) and (6.2.7) as

$$|DR_1(x, y)| \le 3C_1' h^2 (\max_{\substack{(x, y) \in \tau \\ |m|=2}} |D^m u|) (\max_{\substack{(x, y) \in \tau \\ i=1,2,3}} |D\xi_i|) \qquad (6.3.2)$$

We assume here that the derivatives of the shape functions ξ_i satisfy

$$\max_{\substack{(x, y) \in \tau \\ i=1,2,3}} |D\xi_i| \le \frac{C}{h} \qquad (6.3.3)$$

A detailed discussion of this assumption is left for Sec. 6.5. We assume again that C does not depend on τ. The derivative of ξ_i is of order $1/h$ because the height of ξ_i itself is unity while the length of its base is of order h. Then (6.3.2) gives

$$|DR_1(x, y)| \le C_0'' h (\max_{\substack{(x, y) \in \tau \\ |m|=2}} |D^m u|) \qquad (6.3.4)$$

which, together with (6.2.12) and

$$D\epsilon_1(x, y) = Du - D\hat{u}_1 = R_1 - DR_1 \qquad (6.3.5)$$

results in the following error estimate for the derivative of the interpolation:

$$|D\epsilon_1(x, y)| \le C_0 h \max_{\substack{(x, y) \in \tau \\ |m|=2}} |D^m u| \qquad (6.3.6)$$

As mentioned above, we will discuss the inequality (6.3.3) in Sec. 6.5.

6.4 *Error Estimation of the Interpolate Over the Entire Domain*

Once the pointwise error estimate in the triangular element is obtained, it is easy to derive an error estimate for the interpolate $\hat{u}_1$ over the entire domain in terms of the energy norm. The inner product and the norm of the energy space are defined from (4.1.15) by

$$(u, v)_a = a(u, v) = \iint_G \left(\frac{\partial u}{\partial x} \frac{\partial v}{\partial x} + \frac{\partial u}{\partial y} \frac{\partial v}{\partial y} + quv \right) dx \, dy \qquad (6.4.1)$$

$$\|u\|_a = \sqrt{a(u,u)} \tag{6.4.2}$$

It can be proved that the present norm (6.4.2) satisfies the condition (2.2.4) using the Schwarz inequality

$$|(u,v)_a| \le \|u\|_a \|v\|_a \tag{6.4.3}$$

The square of the error of the interpolate in terms of the energy norm is given by

$$\|u - \hat{u}_1\|_{a^2} = a(u - \hat{u}_1, u - \hat{u}_1)$$

$$= \iint_G \left[\left\{ \frac{\partial}{\partial x} (u - \hat{u}_1) \right\}^2 + \left\{ \frac{\partial}{\partial y} (u - \hat{u}_1) \right\}^2 + q(u - \hat{u}_1)^2 \right] dx\, dy \tag{6.4.4}$$

Thus by integrating (6.2.18) and (6.3.6) over the entire domain G we have the following error estimate:

$$\|u - \hat{u}_1\|_a \le Ch \max_{\substack{(x,y) \in G \\ |m|=2}} |D^m u| \tag{6.4.5}$$

where h is the maximum length of the sides of all the triangular elements and the constant C includes the area of the domain G.

6.5 The Uniformity Condition

Now we consider the inequality (6.3.3) that we assumed in the derivation of the error estimate. If the coefficient C on the right-hand side of (6.3.3) differs significantly among different triangular elements, or if it grows rapidly as we repeat mesh refinement, the error of the interpolation may become quite large. In order to avoid these possibilities we assumed in Sec. 6.3 that the triangulation is always performed in such a way that we can choose the coefficient C on the right-hand side of the inequality (6.3.3) so that it does not depend on the triangular element τ. The assumption (6.3.3) is called the *uniformity condition* of the basis functions.

A typical example in which the uniformity condition is violated is the case where a triangular element with an obtuse angle near $180°$ is included. Consider a thin triangular element $P_1 P_2 P_3$ whose largest side $P_2 P_3$ is located parallel to the x axis as shown in Fig. 6.1. Put $\angle P_1 P_2 P_3 = \theta_0$. Then from (5.3.6) we have

$$\frac{\partial \xi_1}{\partial y} = \frac{1}{y_1 - y_2} = \frac{h}{y_1 - y_2} \frac{1}{h} \ge \frac{1}{\tan \theta_0} \frac{1}{h} \tag{6.5.1}$$

Therefore, if mesh refinement is repeated in the process of triangulation in such a way that a triangular element becomes so thin that θ_0 approaches

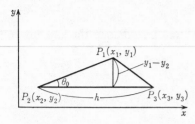

Fig. 6.1 Triangular element for which the uniformity condition may be violated.

zero, $\partial \xi_1 / \partial y$ gradually becomes large and, accordingly, C in (6.3.3) must inevitably be chosen large.

Although a large error of the derivative of the interpolate does not immediately imply a large error of the finite element solution, (6.3.6) or (6.4.5) signifies this possibility, as will be seen in the next section. When applying the FEM, it is important to perform the triangulation in such a way that all the triangular elements do not have an obtuse angle and that each of them is as close as possible to an equilateral triangle.

6.6 *Error of the Finite Element Solution*

From the error estimate of the interpolation obtained above we can derive an error estimate of the finite element solution $\hat{u}_n$ in terms of the energy norm or the Sobolev norm, as in the one-dimensional case discussed in Chap. 3. Since the error estimate (3.4.8) is an abstract result obtained under the assumption that the bilinear form $a(u, v)$ satisfies the ellipticity condition, it can immediately be applied to the problem in two space dimensions, for example, to problem (4.5.4) and (4.5.5). Therefore, if we choose the interpolate $\hat{u}_1$ considered in the preceding sections as $\hat{v}_n$ in (3.4.8), we obtain from (3.4.8) and (6.4.5) an error estimate in terms of the energy norm:

$$\|\hat{u}_n - u\|_a \le Ch \max_{\substack{(x,y) \in G \\ |m|=2}} |D^m u| \qquad (6.6.1)$$

Thus, if the second derivative of the exact solution exists, the error of the finite element solution $\hat{u}_n$ will be of order h provided the triangulation is carried out in such a way that the uniformity condition (6.3.3) is satisfied.

From this estimate and the ellipticity condition (4.3.8) we immediately have an estimate in terms of the Sobolev norm:

$$\|\hat{u}_n - u\|_1 \le C'h \max_{\substack{(x,y) \in G \\ |m|=2}} |D^m u| \qquad (6.6.2)$$

6.7 Numerical Integration and Its Error

Since we have discussed interpolation over a triangle, we take this occasion to give an example of a numerical integration formula over the standard triangle T. Integration over T of the linear interpolate (6.1.2), whose sampling points are the three vertices $(1,0)$, $(0,1)$, and $(0,0)$, results in a numerical integration formula for the integral over T:

$$\iint_T G(\xi_1,\xi_2)\, d\xi_1\, d\xi_2 = \tfrac{1}{6} G(1,0) + \tfrac{1}{6} G(0,1) + \tfrac{1}{6} G(0,0) + \Delta I_T$$

$$(6.7.1)$$

This is an example of (5.8.1). Also, by integrating the error estimate (6.2.18) of the interpolate we have an estimate for the error ΔI_T:

$$|\Delta I_T| \le C \max_{\substack{(\xi_1,\xi_2)\in T \\ |m|=2}} |D^m G| \tag{6.7.2}$$

Transformation of this estimate into the original triangle τ by means of (5.5.3) leads to the expression

$$\iint_\tau g(x,y)\, dx\, dy = \frac{J}{6} g(x_1, y_1) + \frac{J}{6} g(x_2, y_2) + \frac{J}{6} g(x_3, y_3) + \Delta I_\tau$$

$$(6.7.3)$$

where J is the Jacobian of the transformation (5.5.3) and (x_1,y_1), (x_2,y_2), and (x_3,y_3) are the vertices of τ. The error ΔI_τ becomes

$$|\Delta I_T| = |\Delta I_\tau| \le C'' h^2 \max_{\substack{(x,y)\in\tau \\ |m|=2}} |D^m g| \tag{6.7.4}$$

from the relations $G(\xi_1,\xi_2) = J g(x,y)$ and $J = 2S \le C' h^2$.

The error ΔI_τ of the formula (6.7.3) vanishes when $g(x,y)$ is a polynomial of order less than or equal to 1. Generally a numerical integration formula of the type (5.8.3) is said to be the *formula of order* μ if it gives the exact value when $g(x,y)$ is a polynomial of order less than or equal to μ. Hence the formula given above is of order 1.

6.8 Variational Crimes

In the rigorous sense the variational method and the Galerkin method require that the integral (4.1.15) or (4.1.16) be computed exactly. Therefore, if it is computed approximately by means of a numerical integration formula, the result will be beyond the framework of variation. Introducing this kind of approximation beyond the framework of variation is called a *variational crime*. In addition to the application of numerical integration, the introduction of nonconforming basis functions, which will be discussed later, and the approximation of a curved boundary using a polygon are typical examples of variational crimes.

Although variational crimes are committed unavoidably in practice, it is important to control the error due to a variational crime so that it does not become extremely large compared with the error intrinsic to Galerkin's method as estimated by (6.6.2). It is desirable to control the error due to a variational crime to be of the same order of h as the error estimated by (6.6.2).

As an example of a variational crime we consider numerical integration and estimate its contribution to the finite element solution. Let the Galerkin equation considered here be (4.5.4) or, equivalently,

$$a(\hat{u}_n, \hat{v}) = (f, \hat{v}) \qquad \forall \hat{v} \in \mathring{K}_n \tag{6.8.1}$$

where we substitute $q = 0$ into (4.5.4). On the other hand, we write the perturbed equation so that the integration is approximated by means of numerical integration as

$$a_*(\hat{U}_n, \hat{v}) = (f, \hat{v})_* \qquad \forall \hat{v} \in \mathring{K}_n \tag{6.8.2}$$

The subscript $_*$ indicates perturbation, and $a_*(\hat{U}_n, \hat{v})$ and $(f, \hat{v})_*$ in (6.8.2) are defined specifically by

$$a_*(\hat{U}_n, \hat{v}) = \sum_\tau \sum_j A_j \left[\frac{\partial \hat{U}_n}{\partial x} \frac{\partial \hat{v}}{\partial x} + \frac{\partial \hat{U}_n}{\partial y} \frac{\partial \hat{v}}{\partial y} \right]_{x=aj, y=bj} \tag{6.8.3}$$

$$(f, \hat{v})_* = \sum_\tau \sum_j A_j f(a_j, b_j) \hat{v}(a_j, b_j) \tag{6.8.4}$$

where (a_j, b_j) and A_j are the jth point and the corresponding weight of the numerical integration formula over the triangular element τ, respectively. On the boundary both $\hat{u}_n$ and $\hat{U}_n$ vanish.

We assume here that, corresponding to (4.3.8), the bilinear form (6.8.3) also satisfies the ellipticity condition in the sense that

$$\gamma_* \|\hat{v}\|_1^2 \le a_*(\hat{v}, \hat{v}) \qquad \hat{v} \in \mathring{K}_n \tag{6.8.5}$$

where γ_* is a positive constant. If we can choose $\hat{v} \in \mathring{K}_n$ such that both $\partial \hat{v}/\partial x$ and $\partial \hat{v}/\partial y$ vanish at all the sampling points, (6.8.5) does not hold. It is easy to see that such a situation may occur when the number of points in the formula is too small.

As a simple example consider a problem in one space dimension in which a continuous and piecewise quadratic trial function is used. Suppose in this case that a numerical integration formula is used that has only one sampling point in each element, that is, in each subinterval. Then, if we choose $\hat{v} \not\equiv 0$ such that its derivative vanishes at every sampling point, (6.8.5) does not hold. In order to make (6.8.5) valid for this trial function, we must use a formula with at least two points in each subinterval.

6.9 Estimation of the Perturbation Error

In this section we give an error estimate of the perturbed solution $\hat{U}_n$ in an abstract form so that it applies to variational crimes other than numerical integration.

First, note that the following relation holds for any $\hat{v} \in \mathring{K}_n$ from (6.8.1) and (6.8.2):

$$a_*(\hat{u}_n - \hat{U}_n, \hat{v}) = a_*(\hat{u}_n, \hat{v}) - a_*(\hat{U}_n, \hat{v}) = a_*(\hat{u}_n, \hat{v}) - (f, \hat{v})_*$$

$$= (a_* - a)(\hat{u}_n, \hat{v}) + (f, \hat{v}) - (f, \hat{v})_* \qquad (6.9.1)$$

We use the notation

$$(a_* - a)(\hat{u}_n, \hat{v}) \equiv a_*(\hat{u}_n, \hat{v}) - a(\hat{u}_n, \hat{v}) \qquad (6.9.2)$$

to express the difference between a and the perturbed a_* explicitly. From (6.9.1) and the ellipticity condition (6.8.5) we have

$$\gamma_* \|\hat{u}_n - \hat{U}_n\|_1^2 \le a_*(\hat{u}_n - \hat{U}_n, \hat{u}_n - \hat{U}_n)$$

$$= (a_* - a)(\hat{u}_n, \hat{u}_n - \hat{U}_n) + (f, \hat{u}_n - \hat{U}_n) - (f, \hat{u}_n - \hat{U}_n)_* \qquad (6.9.3)$$

Division by $\gamma_* \|\hat{u}_n - \hat{U}_n\|_1$ leads to

$$\|\hat{u}_n - \hat{U}_n\|_1$$

$$\le \frac{1}{\gamma_*} \left\{ \frac{|(a_* - a)(\hat{u}_n, \hat{u}_n - \hat{U}_n)|}{\|\hat{u}_n - \hat{U}_n\|_1} + \frac{|(f, \hat{u}_n - \hat{U}_n) - (f, \hat{u}_n - \hat{U}_n)_*|}{\|\hat{u}_n - \hat{U}_n\|_1} \right\} \qquad (6.9.4)$$

Although if we replace the particular $\hat{u}_n - \hat{U}_n \in \mathring{K}_n$ by a general $\hat{v} \in \mathring{K}_n$, the right-hand side may become either larger or smaller, at least

$$\|\hat{u}_n - \hat{U}_n\|_1$$

$$\le \frac{1}{\gamma^*} \sup_{\hat{v} \in \mathring{K}_n} \left\{ \frac{|(a_* - a)(\hat{u}_n, \hat{v})|}{\|\hat{v}\|_1} + \frac{|(f, \hat{v}) - (f, \hat{v})_*|}{\|\hat{v}\|_1} \right\} \qquad (6.9.5)$$

holds. Therefore from the triangular inequality including the exact solution u of the given problem

$$\|u - \hat{U}_n\|_1 \le \|u - \hat{u}_n\|_1 + \|\hat{u}_n - \hat{U}_n\|_1 \qquad (6.9.6)$$

we eventually obtain the following error estimate for $\hat{U}_n$:

$$\|u - \hat{U}_n\|_1 \leq \|u - \hat{u}_n\|_1$$

$$+ \frac{1}{\gamma*} \sup_{\hat{v} \in \hat{K}_n} \frac{|(a - a_*)(\hat{u}_n, \hat{v})|}{\|\hat{v}\|_1} + \frac{1}{\gamma*} \sup_{\hat{v} \in \hat{K}_n} \frac{|(f, \hat{v}) - (f, \hat{v})_*|}{\|\hat{v}\|_1}$$

$$(6.9.7)$$

This estimate commonly applies to problems whose perturbed equation has the form of (6.8.2).

6.10 *Error of the Numerical Integration of* $a(\hat{u}_n, \hat{v})$

In case of problem (6.8.1) we have already seen in (6.6.2) that the first term on the right-hand side of (6.9.7), that is, the error intrinsic to the Galerkin method, is of order h. Now we explicitly give an estimate of the error induced by numerical integration. We assume here that the numerical integration formula is of order μ; that is, it gives the exact value of the integral when the integrand is a polynomial of order less than or equal to μ.

In this section we obtain an estimate of the second term on the right-hand side of (6.9.7). The numerator of this term becomes

$$(a - a_*)(\hat{u}_n, \hat{v})$$

$$= \iint_G \left(\frac{\partial \hat{u}_n}{\partial x} \frac{\partial \hat{v}}{\partial x} + \frac{\partial \hat{u}_n}{\partial y} \frac{\partial \hat{v}}{\partial y} \right) dx\, dy$$

$$- \sum_\tau \sum_j A_j \left[\left(\frac{\partial \hat{u}_n}{\partial x} \frac{\partial \hat{v}}{\partial x} + \frac{\partial \hat{u}_n}{\partial y} \frac{\partial \hat{v}}{\partial y} \right) \right]_{x=aj,\, y=bj}$$

$$= \sum_\tau \iint_\tau \left\{ \left(\frac{\partial \hat{u}_n}{\partial x} - p_\mu \right) \frac{\partial \hat{v}}{\partial x} + \left(\frac{\partial \hat{u}_n}{\partial y} - q_\mu \right) \frac{\partial \hat{v}}{\partial y} \right\} dx\, dy$$

$$- \sum_\tau \sum_j A_j \left[\left(\frac{\partial \hat{u}_n}{\partial x} - p_\mu \right) \frac{\partial \hat{v}}{\partial x} + \left(\frac{\partial \hat{u}_n}{\partial y} - q_\mu \right) \frac{\partial \hat{v}}{\partial y} \right]_{x=aj,\, y=bj}$$

$$(6.10.1)$$

where p_μ and q_μ are polynomials of order μ defined in each of the triangular elements τ. In (6.10.1) we used the fact that each of these polynomials multiplied by the derivative of $\hat{v}$, that is, by a constant, is integrated exactly by the present formula of order μ. Now we expand $\partial \hat{u}_n/\partial x$ in terms of the Taylor expansion around a point inside the triangular element τ and choose for p_μ the sum of the terms of order up to μ. Then the error is estimated in the same way in which we obtained (6.2.7):

$$\left|\frac{\partial \hat{u}_n}{\partial x} - p_\mu\right| \leq C_1'' h^{\mu+1} \max_{\substack{(x,y)\in\tau \\ |m|=\mu+1}} \left|D^m \frac{\partial \hat{u}_n}{\partial x}\right| \leq C_1'' h^{\mu+1} \max_{\substack{(x,y)\in\tau \\ |m|=\mu+2}} |D^m \hat{u}_n|$$

(6.10.2)

Similarly, we have an estimate for $|\partial \hat{u}_n/\partial y - q_\mu|$.

In the present example, since the piecewise linear basis function $\hat{\phi}_j$ is chosen as $\hat{v}$, $\partial \hat{v}/\partial x$ and $\partial \hat{v}/\partial y$ are actually constants in each τ. Therefore, formally we have

$$|D\hat{v}|^2 = \frac{1}{|S|}\iint_\tau (D\hat{v})^2 \, dx \, dy \leq \frac{1}{|S|}\|\hat{v}\|_{1.\tau}^2$$

(6.10.3)

where the norm in the last term is defined in the triangular element τ and $|S|$ is the area of τ. From this, and applying the Schwarz inequality to (6.10.1), we have

$$|(a - a_*)(\hat{u}_n, \hat{v})| \leq C_1' h^{\mu+1} \|\hat{v}\|_1 \max_{\substack{(x,y)\in G \\ |m|=\mu+2}} |D^m \hat{u}_n|$$

(6.10.4)

Hence the second term on the right-hand side of (6.9.7) is estimated as

$$\frac{1}{\gamma_*} \sup_{\hat{v}\in\hat{K}_n} \frac{|(a_*^- - a)(\hat{u}_n, \hat{v})|}{\|\hat{v}\|_1} \leq C_1 h^{\mu+1} \max_{\substack{(x,y)\in G \\ |m|=\mu+2}} |D^m \hat{u}_n|$$

(6.10.5)

This result shows that, in the present problem, if we use a formula of order 0, that is, a simple formula that integrates exactly only a constant function, the error will be of order h. In addition, since in the present case $\hat{u}_n$ is a piecewise linear function so that $|D^2 \hat{u}_n| = 0$, a formula of order 0 does not give rise to any error at all. This may be evident without the error analysis given above. The purpose of this section is to present a method of error analysis that also applies to more general cases with basis functions of higher order.

6.11 *Error of the Numerical Integration of $(f, \hat{v})$*

In this section we obtain an estimate of the third term on the right-hand side of (6.9.7). Suppose that a formula of order μ is applied. Then the numerator of this term becomes

$$(f, \hat{v}) - (f, \hat{v})_* = \iint_G f\hat{v} \, dx \, dy - \sum_\tau \sum_j A_j f\hat{v}(a_j, b_j)$$

$$= \sum_\tau \iint_\tau (f\hat{v} - r_\mu) \, dx \, dy$$

$$\quad - \sum_\tau \sum_j A_j \{f\hat{v}(a_j, b_j) - r_\mu(a_j, b_j)\}$$

(6.11.1)

where r_μ is a polynomial of order μ defined in each τ. We choose as r_μ the sum of the terms of order up to μ of the Taylor expansion of $f\hat{v}$ in each triangular element τ. Then the error can be expressed as

$$|f\hat{v} - r_\mu| \le C_2'''h^{\mu+1} \max_{\substack{(x,y) \in \tau \\ |m|=\mu+1}} |D^m(f\hat{v})| \qquad (6.11.2)$$

where we assume sufficient differentiability required of f.

When we perform the differentiation $D^m(f\hat{v})$ on the right-hand side of (6.11.2) using Leibniz's formula, the derivatives of $\hat{v}$ that do not vanish are only of order 0 and 1 because $\hat{v}$ is a polynomial of first order. Therefore we have

$$C_2'''h^{\mu+1} \max_{\substack{(x,y) \in \tau \\ |m|=\mu+1}} |D^m(f\hat{v})| \le C_2''h^{\mu+1} (\max_{\substack{(x,y) \in \tau \\ |m| \le \mu+1}} |D^m f|) (\max_{\substack{(x,y) \in \tau \\ |n| \le 1}} |D^n\hat{v}|)$$

$$(6.11.3)$$

Since $D\hat{v}$ is a constant in the present case, (6.10.3) also holds formally. Therefore, even if we integrate (6.11.2) over τ exactly or approximately and sum up the results over the entire domain, the order of h on the right-hand side does not change. Thus from (6.11.1) and (6.11.3) we have

$$|(f, \hat{v}) - (f, \hat{v})_*| \le C_2'h^{\mu+1}\|\hat{v}\|_1 (\max_{\substack{(x,y) \in G \\ |m| \le \mu+1}} |D^m f|) \qquad (6.11.4)$$

and the third term on the right-hand side of (6.9.7) is estimated as

$$\frac{1}{\gamma_*} \sup_{\hat{v} \in \mathring{K}_n} \frac{|(f, \hat{v}) - (f, \hat{v})_*|}{\|\hat{v}\|_1} \le C_2 h^{\mu+1} \max_{\substack{(x,y) \in G \\ |m| \le \mu+1}} |D^m f| \qquad (6.11.5)$$

From (6.11.5) we see that, in order to obtain the result with an error of the same order as that due to the Galerkin approximation, we may use a formula of order $\mu = 0$. Thus, as a whole, by using a numerical integration formula of order 0 we can make the order of the error introduced by a variational crime due to numerical integration equal to the error intrinsic to Galerkin's method.

We used the inequality (6.10.3) in the error analysis presented above. In an error analysis in which a higher-order trial function is used as $\hat{v}$, it is necessary to use, together with an inequality of the form (6.10.2) or (6.11.2), an inequality of the form (6.10.3) for $\hat{v}$ of higher order.

7

Problems With High-Order
Derivatives and
Nonconforming Elements

7.1 *A Differential Equation of the Fourth Order*

The boundary value problems considered so far have involved a differential equation of the second order. In this section we take as an example the following differential equation of the fourth order given in a bounded domain G on the xy plane and show how to deal with differential equations of high order:

$$\Delta^2 u = f \tag{7.1.1}$$

$$u = 0 \quad \text{and} \quad \frac{\partial u}{\partial n} = 0 \quad \text{on } \partial G \tag{7.1.2}$$

where Δ^2 is the *biharmonic operator* defined by

$$\Delta^2 u = \Delta(\Delta u) = \frac{\partial^4 u}{\partial x^4} + 2\frac{\partial^4 u}{\partial x^2 \partial y^2} + \frac{\partial^4 u}{\partial y^4} \tag{7.1.3}$$

This equation describes the displacement of a thin plate bent by a force f. The boundary condition (7.1.2) corresponds to a clamped plate.

In order to apply the FEM to this problem, we must derive a weak form corresponding to (7.1.1) and (7.1.2). In a second-order problem the corresponding weak form includes first-order derivatives, just half of the second, and in order to obtain the weak form we multiply both sides of the original equation by a function $v \in H_1$ or $v \in \mathring{H}_1$ and integrate by parts. Corresponding to this we need to derive a weak form including second-order derivatives, half of the fourth, in the case of a fourth-order problem. To this end we first introduce the Sobolev space H_2 consisting of functions having second-order derivatives in the domain G. The norm is defined by

$$\|v\|_2 = \iint_G \left\{ \left(\frac{\partial^2 v}{\partial x^2}\right)^2 + \left(\frac{\partial^2 v}{\partial x\,\partial y}\right)^2 + \left(\frac{\partial^2 v}{\partial y^2}\right)^2 + \left(\frac{\partial v}{\partial x}\right)^2 + \left(\frac{\partial v}{\partial y}\right)^2 + v^2 \right\} dx\,dy \tag{7.1.4}$$

Moreover, we denote by $\mathring{H}_2$ the subspace of H_2 consisting of functions satisfying

$$v = 0 \qquad \text{and} \qquad \frac{\partial y}{\partial n} = 0 \qquad \text{on } \partial G \tag{7.1.5}$$

Multiplying both sides of the given equation (7.1.1) by $v \in \mathring{H}_2$ and integrating over the domain by parts using (4.1.9) twice, we have

$$\iint_G (\Delta^2 u - f)v\,dx\,dy = \iint_G (\Delta u\,\Delta v - fv)\,dx\,dy$$
$$- \int_{\partial G} \Delta u\,\frac{\partial v}{\partial n}\,d\sigma + \int_{\partial G} \left(\frac{\partial}{\partial n}\,\Delta u\right)v\,d\sigma = 0 \tag{7.1.6}$$

From (7.1.5) the second and third terms on the right-hand side vanish, which results in the weak form

$$\iint_G (\Delta u\,\Delta v - fv)\,dx\,dy = 0 \qquad \forall v \in \mathring{H}_2 \tag{7.1.7}$$

It can be seen from the derivation given above that in general if the original differential equation is of order $2m$, then the weak form includes mth-order derivatives.

7.2 Quadratic Basis Functions

We search for a finite element solution of the form

$$\hat{u}_n(x, y) = \sum_{j=1}^{n} a_j \hat{\varphi}_j(x, y) \tag{7.2.1}$$

as before. It must be noted here, however, that twice-differentiability is required for u in (7.1.7), hence in this problem the basis function $\hat{\phi}_j$ must be such that its second-order derivative is square-integrable, that is, $\hat{\phi}_j \in H_2$.

If a linear function is used on the triangular element, its second-order derivative is identically equal to zero and, in addition, the second-order derivative does not exist along the side common to two adjacent triangular elements. Hence the piecewise linear function used in the preceding chapters cannot be employed in the present problem.

Another possibility is to use piecewise quadratic functions. Suppose that the finite element solution in the triangular element τ is expressed as

$$\hat{u}_n(x,y)|_\tau = c_1 + c_2 x + c_3 y + c_4 x^2 + c_5 xy + c_6 y^2 \qquad (7.2.2)$$

In order to determine the six unknowns $c_j, j = 1, 2, \ldots, 6$, we choose for the nodal points the three vertices P_1, P_2, and P_3 and the three midpoints P_4, P_5, and P_6 of the sides of the triangular element τ, as shown in Fig. 7.1. The six unknowns are determined by specifying the value of $\hat{u}_n$ at these six points.

To compute the matrix elements we use the quadratic shape functions defined in Sec. 5.4. Likewise, for each of the other triangular elements we can determine a piecewise quadratic function.

We can see that the piecewise quadratic function $\hat{u}_n(x,y)$ thus determined is continuous over the entire domain as follows. Since it is obvious that $\hat{u}_n$ is continuous inside each triangular element, the question is whether or not it is continuous along the boundaries between two triangular elements. Let the common sides of the triangular elements τ and τ' be $P_1 P_2$. Since the side $P_1 P_2$ is a line segment, it can be expressed by a linear equation as

$$\alpha x + \beta y + \gamma = 0 \qquad (7.2.3)$$

If we take the limit of $\hat{u}_n$ to $P_1 P_2$ in τ, then $\hat{u}_n$ is obviously given by (7.2.2) along $P_1 P_2$. By eliminating either x or y from (7.2.2) using (7.2.3), we can express $\hat{u}_n$ along $P_1 P_2$ as a quadratic function of either y or x. Suppose that along $P_1 P_2$ this becomes

$$\hat{u}_n(x, y) = d_0 + d_1 x + d_2 x^2 \qquad (7.2.4)$$

where d_0, d_1, and d_2 can be determined uniquely from the data at the three points P_1, P_6, and P_2. On the other hand, if we take the limit of $\hat{u}_n$ to $P_1 P_2$ in τ', then $\hat{u}_n$ can be expressed as a quadratic function of, say x, and is also

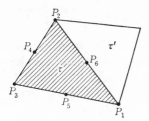

Fig. 7.1 Nodes for a piecewise quadratic basis function.

uniquely determined from the data at P_1, P_6, and P_2. Apparently it coincides with (7.2.4). Thus we conclude that $\hat{u}_n(x,y)$ is continuous along the sides of the triangular elements.

However, the first-order derivative of $\hat{u}_n$ is not continuous along the sides, hence the second-order derivative of $\hat{u}_n$ is not square-integrable over entire domain. That is, a piecewise quadratic function is not an admissible function of the weak form including the second-order derivatives. In order to make the first-order derivative of the trial function continuous, it is necessary to employ higher-order basis functions. Although such basis functions may be constructed in various forms, numerical treatment becomes more complicated as the order number becomes higher.

7.3 Nonconforming Elements

For simplicity we use the piecewise quadratic function described in the preceding section, ignoring the discontinuity of the derivatives along the sides of each triangular element. Then the trial function will not be an admissible function of the given problem in the rigorous sense. The basis functions in terms of which such a trial function is expressed are called *nonconforming elements* in the FEM. The basis functions in terms of which an admissible trial function is expressed are called *conforming elements*. The introduction of nonconforming elements is also a variational crime. In practice, nonconforming elements are often used in order to simplify treatment of the given problem and in many cases lead to a significant result. In this section, for simplicity, we again take a differential equation of second order and consider the effect of employing nonconforming elements.

Consider problem (4.1.13) with $q = 0$ defined in a polygonal domain in two space dimensions:

$$a(u,v) - (f,v) = 0 \qquad \forall v \in \mathring{H}_1 \tag{7.3.1}$$

$$u = 0 \qquad \text{on the boundary } \partial G \tag{7.3.2}$$

where

$$a(u,v) = \iint_G \left(\frac{\partial u}{\partial x}\frac{\partial v}{\partial x} + \frac{\partial u}{\partial y}\frac{\partial v}{\partial y} \right) dx\, dy \tag{7.3.3}$$

We construct here an approximate solution $\hat{U}_n(x,y)$ using piecewise linear functions matched at the midpoint of each side of the triangular element τ as shown in Fig. 7.2. Since this function is not generally continuous along the sides except at the midpoints, it is not an admissible function of (7.3.1).

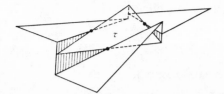

Fig. 7.2 Piecewise linear functions matched at the midpoints of the sides.

Ignoring the continuity along the side between elements implies the use of

$$a_*(u,v) = \sum_\tau \iint_\tau \left(\frac{\partial u}{\partial x} \frac{\partial v}{\partial x} + \frac{\partial u}{\partial y} \frac{\partial v}{\partial y} \right) dx \, dy \qquad (7.3.4)$$

instead of $a(u,v)$ in (7.3.1). Corresponding to this bilinear form we define a seminorm

$$\|v\|_* = [a_*(v,v)]^{1/2} \qquad (7.3.5)$$

We denote as $\mathring{V}_*$ the set of all functions such that the seminorm (7.3.5) is bounded, that they are matched at the midpoint of each side of the elements, and that the value at the midpoint of each side on the boundary ∂G of the domain G is zero. Then the problem can be formulated as

$$a_*(\hat{U}_n, \hat{v}) - (f, \hat{v}) = 0 \qquad \forall \hat{v} \in \mathring{V}_* \qquad (7.3.6)$$

where $\hat{U}_n$ is also an element of $\mathring{V}_*$.

We prove here that with respect to the seminorm (7.3.5)

$$|a_*(u,v)| \le \|u\|_* \|v\|_* \qquad (7.3.7)$$

holds. First it is easily shown, in a way similar to that used in Sec. 3.2, that

$$|a_*(u,v)| \le \sum_\tau \iint_\tau \left(\left| \frac{\partial u}{\partial x} \frac{\partial v}{\partial x} \right| + \left| \frac{\partial u}{\partial y} \frac{\partial v}{\partial y} \right| \right) dx \, dy$$

$$\le \sum_\tau \iint_\tau \left\{ \left(\frac{\partial u}{\partial x} \right)^2 + \left(\frac{\partial u}{\partial y} \right)^2 \right\}^{1/2} \left\{ \left(\frac{\partial v}{\partial x} \right)^2 + \left(\frac{\partial v}{\partial y} \right)^2 \right\}^{1/2} dx \, dy$$

$$\le \sum_\tau \left[\iint_\tau \left\{ \left(\frac{\partial u}{\partial x} \right)^2 + \left(\frac{\partial u}{\partial y} \right)^2 \right\} dx \, dy \right]^{1/2}$$

$$\left[\iint_\tau \left\{ \left(\frac{\partial v}{\partial x} \right)^2 + \left(\frac{\partial v}{\partial y} \right)^2 \right\} dx \, dy \right]^{1/2} \qquad (7.3.8)$$

Note that from the identity

$$\left(\sum_{k=1}^n a_k b_k \right)^2 = \left(\sum_{k=1}^n a_k^2 \right) \left(\sum_{k=1}^n b_k^2 \right) - \frac{1}{2} \sum_{i=1}^n \sum_{j=1}^n (a_i b_j - a_j b_i)^2$$

$$(7.3.9)$$

we immediately have the Cauchy-Schwarz inequality

$$\left(\sum_{k=1}^{n} a_k b_k\right)^2 \leq \left(\sum_{k=1}^{n} a_{k^2}\right)\left(\sum_{k=1}^{n} b_{k^2}\right) \tag{7.3.10}$$

From this expression and from (7.3.8) we eventually have

$$|a_*(u, v)| \leq \left[\sum_\tau \iint_\tau \left\{\left(\frac{\partial u}{\partial x}\right)^2 + \left(\frac{\partial u}{\partial y}\right)^2\right\} dx\, dy\right]^{1/2}$$

$$\left[\sum_\tau \iint_\tau \left\{\left(\frac{\partial v}{\partial x}\right)^2 + \left(\frac{\partial v}{\partial y}\right)^2\right\} dx\, dy\right]^{1/2}$$

$$\tag{7.3.11}$$

that is, (7.3.7).

7.4 *Perturbation Error*

In this section we express the perturbation error introduced by employing the nonconforming elements given above in terms of the energy semi-norm. In the present case the relation corresponding to the ellipticity condition in the H_1 norm becomes an equality:

$$\|\hat{v}\|_*^2 = a_*(\hat{v}, \hat{v}) \qquad \forall \hat{v} \in \mathring{V}_* \tag{7.4.1}$$

from the definition (7.3.5). If we replace $\hat{v}$ by $\hat{w} - \hat{U}_n$, we have

$$\|\hat{w} - \hat{U}_n\|_*^2 = a_*(\hat{w} - \hat{U}_n, \hat{w} - \hat{U}_n)$$

$$= a_*(\hat{w} - u, \hat{w} - \hat{U}_n) + a_*(u, \hat{w} - \hat{U}_n) - a_*(\hat{U}_n, \hat{w} - \hat{U}_n)$$

$$= a_*(\hat{w} - u, \hat{w} - \hat{U}_n) + a_*(u, \hat{w} - \hat{U}_n) - (f, \hat{w} - \hat{U}_n)$$

$$\forall \hat{w} \in \mathring{V}^* $$

$$\tag{7.4.2}$$

where u is the solution of (7.3.1) and (7.3.2) and $\hat{U}_n$ is the solution of (7.3.6). Next we use the inequality (7.3.7) for (7.4.2) and divide it by $\|\hat{U}_n - \hat{w}\|_*$, which leads to

$$\|\hat{w} - \hat{U}_n\|_* \leq \|\hat{w} - u\|_* + \frac{|a_*(u, \hat{w} - \hat{U}_n) - (f, \hat{w} - \hat{U}_n)|}{\|\hat{w} - \hat{U}_n\|_*}$$

$$\leq \|u - \hat{w}\|_* + \sup_{v \in V_*} \frac{|a_*(u, \hat{v}) - (f, \hat{v})|}{\|\hat{v}\|_*} \qquad \forall \hat{w} \in \mathring{V}_*$$

$$\tag{7.4.3}$$

Here we choose for $\hat{w}$ the finite element solution $\hat{u}_n$ expressed in terms of the conforming elements. Then we have

$$\|\hat{u}_n - \hat{U}_n\|_* \le \|u - \hat{u}_n\|_* + \sup_{\hat{v} \in V_*} \frac{|a_*(u, \hat{v}) - (f, \hat{v})|}{\|\hat{v}\|_*} \qquad (7.4.4)$$

Moreover, from the inequality

$$\|u - \hat{U}_n\|_* \le \|u - \hat{u}_n\|_* + \|\hat{u}_n - \hat{U}_n\|_* \qquad (7.4.5)$$

which, like the proper norm, also holds for the seminorm $\|v\|_*$, we obtain an estimate of the error induced when the nonconforming function $\hat{U}_n$ is used:

$$\|u - \hat{U}_n\|_* \le 2\|u - \hat{u}_n\|_* + \sup_{\hat{v} \in V_*} \frac{|a_*(u, \hat{v}) - (f, \hat{v})|}{\|\hat{v}\|_*} \qquad (7.4.6)$$

The difference between the derivation of this estimate and that of (6.9.7) is that in the present case we need to deal with functions in a wider space by introducing the new seminorm (7.3.5), since the trial function $\hat{U}_n$ here does not generally belong to $\mathring{K}_n$. An approximation corresponding to the perturbation in a space consisting of all admissible functions is called an *inner approximation*, while an approximation corresponding to the perturbation in a wider space to which nonadmissible functions may belong is called an *external approximation*. Numerical integration is an example of an inner approximation, while the introduction of nonconforming elements is an example of an external approximation.

7.5 Estimation of Error Arising From a Discrepancy Along the Side

As mentioned previously, if $v \in \mathring{H}_1$, then there is no difference between $\|v\|_a = \{a(v, v)\}^{1/2}$ and $\|v\|_*$. Hence the first term on the right-hand side of (7.4.6) is equal to $2\|u - \hat{u}_n\|_a$. From (6.6.1) we already know that this term is of order h in the present case. Now we obtain an estimate for the second term on the right-hand side of (7.4.6).

Let u be the exact solution of (7.3.1) and (7.3.2) and $\hat{v}$ be any element of $\mathring{V}_*$. Then the application of Green's formula (4.1.9) to (7.3.4) in each triangular element results in

$$\begin{aligned}
a_*(u, \hat{v}) &= -\sum_{\tau} \iint_{\tau} \hat{v} \, \Delta u \, dx \, dy + \sum_{\tau} \int_{\partial \tau} \frac{\partial u}{\partial n} \hat{v} \, d\sigma \\
&= -\iint_{G} \hat{v} \, \Delta u \, dx \, dy + \sum_{\tau} \int_{\partial \tau} \frac{\partial u}{\partial n} \hat{v} \, d\sigma \qquad (7.5.1) \\
&= (f, \hat{v}) + \sum_{\tau} \int_{\partial \tau} \frac{\partial u}{\partial n} \hat{v} \, d\sigma
\end{aligned}$$

where $\partial\tau$ represents the three sides of the triangular element τ. Therefore, if we rewrite the second term on the right-hand side as a sum with respect to each side l of the triangular element, we have

$$a_*(u, \hat{v}) - (f, \hat{v}) = \sum_l \int_l \frac{\partial u}{\partial n} ([\hat{v}]_1 - [\hat{v}]_2) \, d\sigma \qquad (7.5.2)$$

where $[\ ]_1$ and $[\ ]_2$ denote the limits of the common side l in the triangular elements τ_1 and τ_2, respectively. The outward normal n is taken from τ_1 to τ_2.

Since the nonconforming function $\hat{v}$ is linear in the triangular element and is matched at the midpoint of each side, we have

$$\int_l ([\hat{v}]_1 - [\hat{v}]^2) \, d\sigma = 0 \qquad (7.5.3)$$

hence, if p_0 is a zeroth-order polynomial, that is, a constant, (7.5.2) becomes

$$a_*(u, \hat{v}) - (f, \hat{v}) = \sum_l \int_l \left(\frac{\partial u}{\partial n} - p_0 \right) ([\hat{v}]_1 - [\hat{v}]_2) \, d\sigma \qquad (7.5.4)$$

Here we choose for p_0 the first term of the Taylor expansion of $\partial u/\partial n$ around the midpoint of the side l. Then we have

$$\left| \frac{\partial u}{\partial n} - p_0 \right| \le C'h \max_{\substack{(x, y) \in \tau_1, \tau_2 \\ |m|=2}} |D^m u| \qquad (7.5.5)$$

where we assume that C' is a constant independent of the choice of triangular element. Therefore, from (7.5.4) and using the Schwarz inequality (3.1.14) we obtain

$$|a_*(u, \hat{v}) - (f, \hat{v})| \le C'h \{ \max_{\substack{(x, y) \in G \\ |m|=2}} |D^m u| \} \sum_l \left[\int_l |[\hat{v}]_1 - [\hat{v}]_2|^2 \, d\sigma \right]^{1/2}$$

$$(7.5.6)$$

where h is the maximum length of the sides of the triangular elements.

Here we claim that if $\hat{v}$ is an element of $\overset{\circ}{V}_*$ the last term of (7.5.6) is bounded:

$$\int_l |[\hat{v}]_1 - [\hat{v}]_2|^2 \, d\sigma \le C''h \left[\iint_{\tau_1} \left[\left(\frac{\partial \hat{v}}{\partial x} \right)^2 + \left(\frac{\partial \hat{v}}{\partial y} \right)^2 \right] dx \, dy \right.$$

$$+ \iint_{\tau_2} \left[\left(\frac{\partial \hat{v}}{\partial x} \right)^2 + \left(\frac{\partial \hat{v}}{\partial y} \right)^2 \right] dx \, dy \qquad (7.5.7)$$

In order to prove this we introduce a linear transformation composed of rotation and displacement:

$$x = x' \cos \theta - y' \sin \theta + x_0$$
$$y = x' \sin \theta + y' \cos \theta + y_0$$

(7.5.8)

where (x_0, y_0) are the coordinates of the midpoint of the common side l of τ_1 and τ_2. By means of this transformation the two triangular elements τ_1 and τ_2 in xy coordinates are mapped onto two triangles τ_1' and τ_2' in $x'y'$ coordinates without changing the shapes and the areas, and the common side l between τ_1 and τ_2 is mapped onto a line segment l' on the x' axis. The midpoint of l' coincides with the origin of the $x'y'$ plane. The Jacobian of this transformation is unity, and in particular $d\sigma = dx'$. Taking into account these relations, we transform (7.5.7) and obtain the following inequality having the same form as (7.5.7):

$$\int_{l'} |[\hat{v}]_1 - [\hat{v}]_2|^2 \, dx' \leq C''h \left[\iint_{\tau_1'} \left[\left(\frac{\partial \hat{v}}{\partial x'} \right)^2 + \left(\frac{\partial \hat{v}}{\partial y'} \right)^2 \right] dx' \, dy' \right.$$
$$\left. + \iint_{\tau_2'} \left[\left(\frac{\partial \hat{v}}{\partial x'} \right)^2 + \left(\frac{\partial \hat{v}}{\partial y'} \right)^2 \right] dx' \, dy' \right]$$

(7.5.9)

Therefore in order to prove the inequality (7.5.7) it is sufficient to prove the case where the common side of the triangular elements τ_1 and τ_2 is located on the x axis and the midpoint coincides with the origin, as shown in Fig. 7.3.

Let the length of side l be h_0. Assume that all the triangular elements have areas of almost the same order of magnitude and that they are not extremely thin. That is, in regard to the area S_τ of each triangular element τ, we assume that

$$C_0 h^2 \leq S_\tau$$

(7.5.10)

where C_0 is a constant independent of h. A linear function $\hat{v}$ in τ_1 and in τ_2

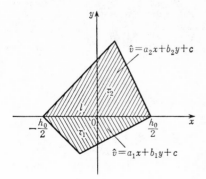

Fig. 7.3 Linear functions in adjacent triangular elements.

can be written as

$$\hat{v} = \begin{cases} a_1 x + b_1 y + c & \text{in } \tau_1 \\ a_2 x + b_2 y + c & \text{in } \tau_2 \end{cases} \qquad (7.5.11)$$

Since the midpoint of l coincides with the origin, the constant terms are set equal to each other. Then we have

$$\int_l |[\hat{v}]_1 - [\hat{v}]_2|^2 \, d\sigma = \int_{-h_0/2}^{h_0/2} \{(a_1 - a_2)x\}^2 \, dx = \frac{h_0^3}{12} (a_1 - a_2)^2$$

$$\leq \frac{h_0^3}{6} (a_1^2 + a_2^3) \leq \frac{h^3}{6} (a_1^2 + a_2^2)$$

$$\leq \frac{h^3}{6} \{(a_1^2 + b_1^2) + (a_2^2 + b_2^2)\}$$

$$(7.5.12)$$

On the other hand, if we write the areas of τ_1 and τ_2 as S_1 and S_2, respectively, we obtain

$$\iint_{\tau_1} \left[\left(\frac{\partial \hat{v}}{\partial x}\right)^2 + \left(\frac{\partial \hat{v}}{\partial y}\right)^2 \right] dx \, dy + \iint_{\tau_2} \left[\left(\frac{\partial \hat{v}}{\partial x}\right)^2 + \left(\frac{\partial \hat{v}}{\partial y}\right)^2 \right] dx \, dy$$

$$= S_1(a_1^2 + b_1^2) + S_2(a_2^2 + b_2^2)$$

$$\geq C_0 h^2 [(a_1^2 + b_1^2) + (a_2^2 + b_2^2)]$$

$$(7.5.13)$$

from (7.5.11) and (7.5.10). Therefore, from (7.5.12) and (7.5.13) we have (7.5.7), where $C'' = 1/6C_0$.

In general, when $A_l \geq 0$, we have from (7.3.10)

$$\sum_{l=1}^{L} A_l^{1/2} \leq \sqrt{L} \left\{ \sum_{l=1}^{L} A_l \right\}^{1/2} \qquad (7.5.14)$$

and, if we let L be the total number of sides in the triangular network in the domain, $\sqrt{L}h$ has the same order of magnitude as the diameter of the domain. Thus we conclude that (7.5.6) is estimated by

$$|a_*(u, \hat{v}) - (f, \hat{v})| \leq Ch \{ \max_{\substack{(x,y) \in G \\ |m|=2}} |D^m u| \} \|\hat{v}\|_* \qquad (7.5.15)$$

7.6 Error Due to Nonconforming Elements

From the inequality (7.5.15) we have for any $\hat{v} \in \mathring{V}_*$

$$\frac{|a_*(u, \hat{v}) - (f, \hat{v})|}{\|\hat{v}\|_*} \leq Ch \max_{\substack{(x, y) \in G \\ |m|=2}} |D^m u| \qquad (7.6.1)$$

This gives an estimate of the perturbation error due to the second term on the right-hand side of (7.4.6). We have already seen in the preceding section that $\|u - \hat{u}_n\|_*$ is of order h. Thus we can conclude that the introduction of nonconforming elements in the present problem does not produce a serious effect on the result.

The seminorm (7.3.5) vanishes when v is a constant function, hence a small error estimate in terms of this seminorm does not imply a small error itself. However, the functions in $\mathring{V}_*$ that we employed in the preceding section vanish at the midpoint of each side along the boundary ∂G and, in addition, are matched at the midpoint of each side of the triangular elements. Under these constraints we can conclude that a small error seminorm implies a small error. Thus, in general, when the error is estimated in terms of a seminorm, another constraint with respect to the derivatives lower than the derivative appearing in the seminorm is necessary.

The linear functional introduced in estimation of the perturbation error due to numerical integration or to nonconforming elements vanishes if the function, like its variable, is a polynomial of order 0 or 1. For example, if we regard the right-hand side of (7.5.2) as a linear functional of u

$$F[u] = \sum_l \int_l \frac{\partial u}{\partial n} ([v]_1 - [v]_2) \, d\sigma \qquad (7.6.2)$$

then it vanishes when u is a linear polynomial or when $\partial u/\partial n$ is a constant function p_0. And we have made efficient use of this property. In fact, we applied a strategy to the perturbation error analysis in which we replaced the function in the linear functional by a difference between the function itself and an appropriate polynomial. The theorem that guarantees mathematically a generalization of this strategy is called the *Bramble-Hilbert lemma* and is often used in the estimation of perturbation error.

7.7   The Mixed Method

Another possibility in dealing with a high-order differential equation is to modify it into a system of lower-order differential equations. In this section we consider the following model problem involving a fourth-order differential equation with one variable:

$$\frac{d^4u}{dx^4} = f \qquad (7.7.1)$$

$$u(0) = u(1) = 0 \qquad (7.7.2)$$

$$u''(0) = u''(1) = 0 \qquad (7.7.3)$$

This is the problem of bending a simply supported beam. If we introduce

$$w = \frac{d^2u}{dx^2} \tag{7.7.4}$$

which corresponds to the moment, (7.7.1) becomes

$$\frac{d^2u}{dx^2} - w = 0 \tag{7.7.5}$$

$$\frac{d^2w}{dx^2} = f \tag{7.7.6}$$

and the boundary condition becomes

$$u(0) = u(1) = 0 \tag{7.7.7}$$

$$w(0) = w(1) = 0 \tag{7.7.8}$$

To derive a weak form, we multiply any element v of the function space $\mathring{H}_1$, which is introduced based on (7.7.7) and (7.7.8), and integrate by parts using (7.7.7) and (7.7.8). Then we obtain a system of equations in the weak form:

$$\int_0^1 \left(\frac{du}{dx} \frac{dv}{dx} + wv \right) dx = 0 \qquad \forall v \in \mathring{H}_1 \tag{7.7.9}$$

$$\int_0^1 \left(\frac{dw}{dx} \frac{dv}{dx} + fv \right) dx = 0 \qquad \forall v \in \mathring{H}_1 \tag{7.7.10}$$

The original equation is of fourth order, while this system of equations includes only first-order derivatives; hence in order to solve this system of equations we need to use only piecewise linear basis functions.

We approximate u and w, respectively, in the following form in terms of piecewise linear basis functions (1.3.3):

$$\hat{u}_n(x) = \sum_{j=1}^{n-1} a_j \hat{\varphi}_j(x) \tag{7.7.11}$$

$$\hat{w}_n(x) = \sum_{j=1}^{n-1} b_j \hat{\varphi}_j(x) \tag{7.7.12}$$

Substituting these trial functions into (7.7.9) and (7.7.10), taking $\hat{\phi}_j$, $j = 1, 2, \ldots, n-1$, as v, and integrating, we have a system of linear equations whose coefficient matrix is $2(n-1) \times 2(n-1)$ with unknowns a_j, b_j, $j = 1, 2, \ldots, n-1$. By solving this system of linear equations we obtain a finite element solution.

In the present problem u and w correspond to the displacement and the moment of the beam, respectively. This method is equivalent to a procedure in which a functional is constructed by mixing more than one different physical quantity, and an approximate solution is obtained by making the functional stationary. Thus it is called the *mixed method*, and it will be considered again in Chap. 14.

7.8   *The Stationary Condition of a Functional*

Actually, the system of equations (7.7.9) and (7.7.10) in the weak form can also be obtained by making the functional

$$J[u, w] = -\frac{1}{2} \int_0^1 \left(2\frac{du}{dx}\frac{dw}{dx} + w^2 + 2fu \right) dx \qquad (7.8.1)$$

stationary. In fact, if we substitute $u + \epsilon\eta$ and $w + \epsilon'\zeta$ (where $\eta, \zeta \in \mathring{H}_1$) into u and w in (7.8.1), respectively, we have

$$J[u + \epsilon\eta, w + \epsilon'\zeta] = J[u, w] + \epsilon \int_0^1 \left(\frac{d^2w}{dx^2} - f \right)\eta \, dx$$

$$+ \epsilon' \int_0^1 \left(\frac{d^2u}{dx^2} - w \right)\zeta \, dx - \left\{ \epsilon\epsilon' \int_0^1 \frac{d\eta}{dx}\frac{d\zeta}{dx} \, dx \right.$$

$$\left. + \frac{1}{2}\epsilon'^2 \int_0^1 \zeta^2 \, dx \right\} \qquad (7.8.2)$$

From the condition that the first variation must vanish we immediately obtain (7.7.5) and (7.7.6). However, note that the sign of the second variation is not definite, owing to the term $\epsilon\epsilon'$. Therefore, although the solution (7.7.5) through (7.7.8) is a stationary point of the functional (7.8.1), it is neither the minimum nor the maximum point of (7.8.1). Hence the method of error estimation based on the positive definiteness of the functional, as shown in Chap. 3, does not apply to such a case.

8

The Heat Equation in One Space Dimension

8.1 *Discretization of a Space Variable*

The FEM can also be applied to time-dependent problems such as diffusion problems and wave problems. To see how this method is used we first consider the following one-dimensional heat conduction problem on the unit interval $(0, 1)$:

$$\frac{\partial u(x, t)}{\partial t} = \sigma \frac{\partial^2 u(x, t)}{\partial x^2} \tag{8.1.1}$$

$$u(0, t) = u(1, t) = 0 \tag{8.1.2}$$

$$u(x, 0) = u_0(x) \tag{8.1.3}$$

where $u(x, t)$ is the temperature at time t at point x and σ is the heat diffusion coefficient and is assumed to be constant. For simplicity the temperatures at both ends are assumed to vanish, and $u_0(x)$ is the initial distribution of the temperature.

Since this problem has two independent variables x and t, we might be able to treat it as a two-dimensional problem on the xt plane and apply the FEM for two dimensions as described in the preceding chapters. However, the space variable x and the time variable t are essentially different kinds of physical quantities, hence it is more natural to discretize x and t independently rather than to treat them as quantities with similar properties. Especially for two- and three-dimensional problems, we can formulate a physical phenomenon into a mathematical model more easily in this way.

We begin with discretization of the space variable. This stage is quite similar to that used in stationary problems. We divide the unit interval $(0, 1)$ of x into n equal subintervals, denoting the nodes by

$$x_j = jh \qquad j = 0, 1, \ldots, n \qquad (8.1.4)$$

and construct the basis functions $\hat{\phi}_j(x), j = 1, 2, \ldots, n-1$, as in (1.3.3). Note that the basis functions do not depend on time t. In the next stage we apply a conventional method widely used in solving approximately time-dependent problems. In this method an approximate solution $\hat{u}_n(x, t)$ corresponding to $u(x, t)$ is expanded in terms of $\{\hat{\phi}_j(x)\}$:

$$\hat{u}_n(x, t) = \sum_{j=1}^{n-1} a_j(t) \hat{\varphi}_j(x) \qquad (8.1.5)$$

The time dependency is placed in the coefficient a_j of the expansion. We substitute this expression into (8.1.1) and multiply both sides by $\hat{\phi}_k(x)$. Then, integrating by parts over $(0, 1)$ and taking into account the boundary condition (8.1.2), we obtain a system of equations

$$\sum_{j=1}^{n-1} \left(\int_0^1 \hat{\varphi}_j \hat{\varphi}_k \, dx \right) \frac{da_j}{dt} + \sum_{j=1}^{n-1} \sigma \left(\int_0^1 \frac{d\hat{\varphi}_j}{dx} \frac{d\hat{\varphi}_k}{dx} \, dx \right) a_j = 0$$

$$k = 1, 2, \ldots, n-1 \qquad (8.1.6)$$

We express the initial condition (8.1.3) in the weak form also:

$$\sum_{j=1}^{n-1} \left(\int_0^1 \hat{\varphi}_j \hat{\varphi}_k \, dx \right) a_j(0) = \int_0^1 u_0 \hat{\varphi}_k \, dx \qquad k = 1, 2, \ldots, n-1$$

$$(8.1.7)$$

If we write

$$\boldsymbol{a} = \boldsymbol{a}(t) = \begin{pmatrix} a_1(t) \\ a_2(t) \\ \vdots \\ a_{n-1}(t) \end{pmatrix} \qquad (8.1.8)$$

$$\boldsymbol{a}_0 = \boldsymbol{a}(0) \qquad (8.1.9)$$

then (8.1.6) and (8.1.7) become

$$M \frac{d\boldsymbol{a}}{dt} + K\boldsymbol{a} = 0 \qquad (8.1.10)$$

$$M\boldsymbol{a}_0 = \boldsymbol{u}_0 \qquad (8.1.11)$$

where u_0 is a vector with $n - 1$ entries whose kth entry is

$$\int_0^1 u_0 \hat{\varphi}_k \, dx \qquad (8.1.12)$$

The entries of the matrices M and K can be computed from (1.4.4) and (1.4.5); that is, M and K correspond to the mass matrix and the stiffness matrix, respectively, in stationary problems and can be written explicitly as

$$M = \frac{h}{6} \begin{bmatrix} 4 & 1 & & & & \\ 1 & 4 & 1 & & 0 & \\ & 1 & 4 & & & \\ & & & \ddots & & \\ & 0 & & & 4 & 1 \\ & & & & 1 & 4 \end{bmatrix} \qquad (8.1.13)$$

$$K = \frac{\sigma}{h} \begin{bmatrix} 2 & -1 & & & & \\ -1 & 2 & -1 & & 0 & \\ & -1 & 2 & & & \\ & & & \ddots & & \\ & 0 & & & 2 & -1 \\ & & & & -1 & 2 \end{bmatrix} \qquad (8.1.14)$$

8.2   Discretization of Time

The system of equations (8.1.10) is a system of linear differential equations with respect to $a_j(t), j = 1,2, \ldots, n - 1$. In order to solve it approximately we carry out the discretization of time t.

We fix a small time interval Δt, discretize time t as

$$t_k = k \, \Delta t \qquad k = 0, 1, 2, \ldots \qquad (8.2.1)$$

and replace the time derivative on the left-hand side of (8.1.10) by the *time difference*

$$\frac{da/dt}{[a(t + \Delta t) - a(t)]/\Delta t} \qquad (8.2.2)$$

In this way the system of equations (8.1.10) is replaced by

$$M \frac{a((k+1)\ \Delta t) - a(k\ \Delta t)}{\Delta t} + Ka(k\ \Delta t) = 0 \qquad (8.2.3)$$

which can be written as

$$Ma((k+1)\ \Delta t) = (M - \Delta t\ K)a(k\ \Delta t) \qquad k = 0,1,2,\ldots \tag{8.2.4}$$

The initial value of $a(t)$ is obtained from (8.1.11) as $a_0 = M^{-1}u_0$. We compute the right-hand side $(M - \Delta t K)a(k\ \Delta t)$ of (8.2.4) for $k = 0,1,2,\ldots$ and solve the system of linear equations (8.2.4) for the unknowns $a((k+1)\ \Delta t)$ repeatedly, obtaining an approximate solution (8.1.5).

Although we have chosen $t = k\ \Delta t$ in the second term on the left-hand side of (8.2.3), we may also choose $t = (k+1)\ \Delta t$ as an alternative and solve

$$(M + \Delta t\ K)a((k+1)\ \Delta t) = Ma(k\ \Delta t) \tag{8.2.5}$$

repeatedly. Here we generalize by introducing a parameter θ in order to mix both cases:

$$M \frac{a((k+1)\ dt) - a(k\ \Delta t)}{\Delta t} + K\{\theta a((k+1)\ \Delta t) +$$

$$(1 - \theta)a(k\ \Delta t)\} = 0 \qquad 0 \le \theta \le 1 \tag{8.2.6}$$

which can be written as

$$(M + \theta\ \Delta t\ K)a((k+1)\ \Delta t) = (M - (1 - \theta)\ \Delta t\ K)a(k\ \Delta t) \tag{8.2.7}$$

Scheme (8.2.4) corresponds to the case where $\theta = 0$ and is called the *forward scheme*, and scheme (8.2.5) corresponds to the case where $\theta = 1$ and is called the *backward scheme*. In particular, the scheme for $\theta = \frac{1}{2}$

$$(M + \tfrac{1}{2}\ \Delta t\ K)a((k+1)\ \Delta t) = (M - \tfrac{1}{2}\ \Delta t\ K)a(k\ \Delta t) \tag{8.2.8}$$

is called the *Crank-Nicolson* scheme and is often used in practice.

In the formulation stated above, the initial condition is also transformed into the weak form (8.1.7) in order to treat the problem uniformly from the viewpoint of the weak form. It is easy to see that the coefficients $a_0 = a(0)$ minimize

$$\int_0^1 \left(u_0(x) - \sum_{j=1}^{n-1} a_j(0)\hat\varphi_j(x) \right)^2 dx \tag{8.2.9}$$

There are, however, other ways to treat the initial condition. For example, direct substitution of the initial condition

$$a_j(0) = u_0(x_j) \qquad j = 1, 2, \ldots, n - 1 \qquad (8.2.10)$$

may also be another good choice.

8.3 *Lumped Mass System*

The mass matrix in the present case is tridiagonal. Equation (8.2.7) becomes a system of linear equations not only when $\theta \neq 0$ but also when $\theta = 0$, and we need to solve the system of equations repeatedly at each time t_k. However, if the mass matrix M can be replaced by a diagonal matrix, we need not solve the system of equations and we can carry out the computation with a much smaller number of arithmetic operations.

The integral defining each entry of the mass matrix does not include the derivative of the basis function $\hat{\phi}_k$, so that differentiability of the basis function is not required as far as computation of the mass matrix is concerned. Hence we divide this interval $(0, 1)$ into n subintervals in the same way as in the case of $\hat{\phi}_k$ and introduce a set of simple basis functions $\bar{\phi}_k$ (Fig. 8.1):

$$\bar{\varphi}_k(x) = \begin{cases} 0 & 0 \leq x < \frac{1}{2}(x_{k-1} + x_k) \\ 1 & \frac{1}{2}(x_{k-1} + x_k) \leq x < \frac{1}{2}(x_k + x_{k+1}) \\ 0 & \frac{1}{2}(x_k + x_{k+1}) \leq x \leq 1 \end{cases} \qquad (8.3.1)$$

This basis function is a zeroth-order piecewise polynomial.

The *ij* entry of the mass matrix M based on this set of basis functions becomes

$$\int_0^1 \varphi_i \bar{\varphi}_j \, dx = \begin{cases} h & i = j \\ 0 & i \neq j \end{cases} \qquad (8.3.2)$$

The replacement of $\hat{\phi}_k$ by $\bar{\phi}_k$ in the definition of the mass matrix is called *lumping of the mass*. By lumping the mass matrix becomes

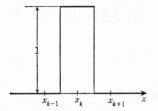

Fig. 8.1 Piecewise constant basis function $\bar{\phi}_k$.

$$M = hI \qquad (8.3.3)$$

where I is the identity matrix. Even if a factor that depends on the space variable is multiplied on the left side of (8.1.1), the mass matrix becomes diagonal if we perform lumping. If the mass matrix is diagonal, all the rows on the left-hand side of (8.2.4) are independent of each other, and (8.2.4) can be solved by means of simple arithmetic division.

The system of equations obtained using the piecewise constant basis functions (8.3.1) in the mass matrix M is called the *lumped mass system*. On the other hand, the system of equations in which piecewise linear polynomials are used in the mass matrix M as in the stiffness matrix K is called the *consistent mass system* because there is a consistency in M and K in the sense that $\hat{\phi}_k$ is used both in M and in K. Note that in the lumped mass system the integrand of the first term on the left-hand side of (8.1.6) is replaced:

$$\hat{\varphi}_j \hat{\varphi}_k \rightarrow \bar{\varphi}_j \bar{\varphi}_k \qquad (8.3.4)$$

and the solution $a(k\,\Delta t)$ of (8.2.4) is substituted on the right-hand side of (8.1.5) to obtain an approximate solution. Therefore the approximate solution in the case of the lumped mass system is also a piecewise linear polynomial.

As stated above, the lumped mass system is, in the first stage, considered to be introduced in order to diagonalize the mass matrix and to avoid solving a system of linear equations. However, as will be mentioned in later sections, the solution obtained with the lumped mass system is generally more stable than that obtained with the consistent mass system. For this reason the lumped mass system is used in (8.2.6) not only for $\theta = 0$ but also for $\theta \neq 0$.

8.4 The Finite Element Method and the Finite Difference Method

As an alternative method for solving (8.1.1) the FEM is well known. In this method we discretize the space variable x with a mesh h, discretize the time variable t with a time step Δt, and replace the derivatives in (8.1.1) by the finite differences. Then we have

$$\frac{u(jh, (k+1)\,\Delta t) - u(jh, k\,\Delta t)}{\Delta t}$$

$$= \sigma \frac{u((j-1)h, k\,\Delta t) - 2u(jh, k\,\Delta t) + u((j+1)h, k\,\Delta t)}{h^2}$$

$$(8.4.1)$$

This is known as the *forward difference scheme*. If we replace $k \, \Delta t$ in u on the right-hand side by $t = (k + 1) \, \Delta t$, we will have the *backward difference scheme*. Multiplying both sides by $h \, \Delta t$, followed by some modification, we have from (8 4.1)

$$M u((k + 1) \, \Delta t) = (M - \Delta t \, K) u(k \, \Delta t) \tag{8.4.2}$$

where $u(t)$ is a vector whose jth entry is $u(jh, t)$, and M and K are matrices defined by (8.3.3) and (8.1.14), respectively.

If we express a finite element solution in the form of (8.1.5), we have

$$\hat{u}_n(jh, t) = a_j(t) \tag{8.4.3}$$

Therefore we see that in the present simple model problem the finite difference scheme (8.4.2) and the finite element scheme (8.2.4) based on the lumped mass system are identical. If the initial value is the same, the solutions obtained with these two schemes coincide at the nodal points.

As stated above, the FEM is closely related to the finite difference method. However, the matrix entry in the FEM is given by an integral in the near neighborhood of a nodal point, and the matrix entry in the finite difference method is given by a value defined at the nodal point. In other words, in the finite difference method an actual value at the nodal point is used directly, while in the FEM an averaged value in the neighborhood of the nodal point is used. Numerical discrepancies due to this difference, however, are usually small.

8.5   Error of the Lumped Mass System

In this section we obtain an estimation of error for the finite element solution

$$a_j^k = a_j(k \, \Delta t) = \hat{u}_n(x_j, k \, \Delta t) \tag{8.5.1}$$

obtained using scheme (8.2.6) or (8.2.7). Let the value of the exact solution of (8.1.1) through (8.1.3) at the node $x = x_j$ and $t = k \, \Delta t$ be

$$u_j^k = u(x_j, k \, \Delta t) \tag{8.5.2}$$

First we consider the lumped mass system. If we multiply $1/h$ by scheme (8.2.6) and rewrite it in terms of (8.3.3), we have

$$\frac{1}{\Delta t} (a_j^{k+1} - a_j^k) + \frac{\sigma}{h^2} \{\theta(-a_{j-1}^{k+1} + 2a_j^{k+1} - a_{j+1}^{k+1})$$

$$+ (1 - \theta)(-a_{j-1}^k + 2a_j^k - a_{j+1}^k)\} = 0$$

$$\tag{8.5.3}$$

We assume here that the exact solution u is sufficiently smooth and expand $u_j{}^{k+1}$, u_j^k, and so on, around $(x_j, (k + \frac{1}{2})\,\Delta t)$ in terms of a two-point Taylor series. Then we obtain

$$\frac{u_j^{k+1} - u_j^k}{\Delta t} = \frac{\partial u}{\partial t}\left(x_j, \left(k + \frac{1}{2}\right)\Delta t\right) + O(\Delta t^2) \tag{8.5.4}$$

$$\theta\left(\frac{-u_{j-1}^{k+1} + 2u_j^{k+1} - u_{j+1}^{k+1}}{h^2}\right) + (1 - \theta)\left(\frac{-u_{j-1}^k + 2u_j^k - u_{j+1}^k}{h^2}\right)$$

$$= -\frac{\partial^2 u}{\partial x^2}\left(x_j, \left(k + \frac{1}{2}\right)\Delta t\right) - \left\{\sigma\left(\theta - \frac{1}{2}\right)\Delta t + \frac{1}{12}h^2\right\}\frac{\partial^4 u}{\partial x^4}\left(x_j, \left(k + \frac{1}{2}\right)\Delta t\right)$$

$$+ O(\Delta t^2) + O(h^2\,\Delta t) + O(h^4) \tag{8.5.5}$$

Here we used (8.1.1) because u satisfies it. Now we write the error of the finite element solution:

$$e_j^k = a_j^k - u_j^k \tag{8.5.6}$$

Note that u_j^k is the exact solution satisfying

$$\frac{\partial u}{\partial t}\left(x_j, \left(k + \frac{1}{2}\right)\Delta t\right) = \sigma\frac{\partial^2 u}{\partial x^2}\left(x_j, \left(k + \frac{1}{2}\right)\Delta t\right) \tag{8.5.7}$$

Then from (8.5.3) through (8.5.5) we see that the error e_j^k satisfies the scheme

$$\frac{1}{\Delta t}(e_j^{k+1} - e_j^k) + \frac{\sigma}{h^2}\{\theta(-e_{j-1}^{k+1} + 2e_j^{k+1} - e_{j+1}^{k+1}) + (1 - \theta)$$

$$(- e_{j-1}^k + 2e_j^k - e_{j+1}^k)\} = r_j^k \tag{8.5.8}$$

where

$$r_j^k = \sigma\left\{\sigma\left(\theta - \frac{1}{2}\right)\Delta t + \frac{1}{12}h^2\right\}\frac{\partial^4 u}{\partial x^4}\left(x_j, \left(k + \frac{1}{2}\right)\Delta t\right)$$

$$+ O(\Delta t^2) + O(h^2\,\Delta t) + O(h^4) \tag{8.5.9}$$

Now we assume that the mesh size h of the space variable and the step size Δt of the time variable always satisfy the relation

$$\frac{\sigma\,\Delta t}{h^2} = \lambda = \text{constant} \tag{8.5.10}$$

Then we have

$$r_j^k = O(\Delta t) \tag{8.5.11}$$

and we see that scheme (8.5.8) satisfied by the error becomes

$$-\theta\lambda e_{j-1}^{k+1} + (1 + 2\theta\lambda)e_j^{k+1} - \theta\lambda e_{j+1}^{k+1}$$

$$= (1 - \theta)\lambda e_{j-1}^k + (1 - 2(1 - \theta)\lambda)e_j^k + (1 - \theta)\lambda e_{j+1}^k + \Delta t\, r_j^k \tag{8.5.12}$$

If we move the first and third terms on the left-hand side to the right-hand side, we have

$$(1 + 2\theta\lambda)\, e_j^{k+1} = \theta\lambda\, e_{j-1}^{k+1} + \theta\lambda\, e_{j+1}^{k+1} + (1-\theta)\lambda\, e_{j-1}^{k} + (1 - 2(1-\theta)\lambda)\, e_j^{k}$$
$$+ (1-\theta)\,\lambda e_{j+1}^{k} + \Delta t\; r_j^{k} \tag{8.5.13}$$

Except for the coefficient of the fourth term on the right-hand side, the coefficients of all the terms on both sides are nonnegative. Thus we assume that the coefficient of the fourth term on the right-hand side is also nonnegative; that is,

$$1 - 2(1-\theta)\lambda \geq 0 \tag{8.5.14}$$

Let

$$\max_{1 \leq j \leq n-1} |e_j^k| = e^{(k)} \tag{8.5.15}$$

$$\max_{j,k} |r_j^k| = r \tag{8.5.16}$$

Note that, since the approximate solution satisfies the exact boundary condition (8.1.2), there is no error on the boundary. Suppose that, at $t = (k + 1)\,\Delta t$, $|e_j^{k+1}|$ attains a maximum for $j = m$ and substitute $j = m$ into (8.5.13). Note that the coefficient of each term of (8.5.13) is nonnegative. Then, taking the absolute values of both sides, we have

$$(1 + 2\theta\lambda)e^{(k+1)} = (1 + 2\theta\lambda)|e_m^{k+1}|$$
$$\leq \theta\lambda|e_{m-1}^{k+1}| + \theta\lambda|e_{m+1}^{k+1}| + (1-\theta)\lambda|e_{m-1}^{k}| + (1 - 2(1-\theta)\lambda)|e_m^{k}|$$
$$+ (1-\theta)\lambda|e_{m+1}^{k}| + \Delta t|r_m^{k}|$$
$$\leq 2\theta\lambda e^{(k+1)} + (1-\theta)\lambda e^{(k)} + (1 - 2(1-\theta)\lambda)e^{(k)} + (1-\theta)\lambda e^{(k)} + \Delta t\, r$$
$$\leq 2\theta\lambda e^{(k+1)} + e^{(k)} + \Delta t\, r$$
$$\tag{8.5.17}$$

This reduces to

$$e^{(k+1)} \leq e^{(k)} + \Delta t\, r \tag{8.5.18}$$

After repeated substitution starting from $k = 0$, we finally have from this inequality

$$e^{(k)} \leq e^{(0)} + (k\,\Delta t)r$$
$$= e^{(0)} + tO(\Delta t) \qquad t = k\,\Delta t \tag{8.5.19}$$

Therefore we see that, if the initial error $e^{(0)}$ is sufficiently small and Δt is made small in such a way that (8.5.10) and (8.5.14) are satisfied, the error of the finite element solution is bounded small.

8.6 Stability of the Lumped Mass System and the Maximum Principle

In the estimating error we assumed (8.5.14). Although this assumption was made in order to establish an upper bound for the error in the preceding section, it is significant from the standpoint of numerical computation and also of physics.

If we assume $\theta = 0$ for simplicity, scheme (8.5.3) becomes

$$a_j^{k+1} = \lambda a_{j-1}^k + (1 - 2\lambda)a_j^k + \lambda a_{j+1}^k \qquad (8.6.1)$$

Suppose that assumption (8.5.14) does not hold. Then, if the temperature at time $t = k \, \Delta t$ is $a_j^k > 0$ at the nodal point $j = J$, and $a_j^k = 0$ at other nodal points $j \ne J$, it turns out that $a_j^{k+1} < 0$ from (8.6.1). This means that, although the temperature is initially positive, it may become negative after one time step. This is contradictory to the properties of the original heat equation. Furthermore, if we repeat the computation for $t = (k + 2) \, \Delta t$, $(k + 3) \, \Delta t$, ..., the temperature distribution turns out to have an extraordinary shape quite different from that of the natural temperature distribution given by (8.1.1) through (8.1.3). In Fig. 8.2 the change in a_j^k for $\lambda = \frac{3}{4}$ and $J = 5$ is shown. As seen above, violation of condition (8.5.14) may lead to great instability in the approximate solution. For this reason condition (8.5.14), that is,

$$\frac{1}{2(1 - \theta)} \geq \frac{\sigma \, \Delta t}{h^2} \qquad (8.6.2)$$

is called the *stability condition* for the scheme.

In general the exact solution $u(x, t)$ of the heat equation (8.1.1) attains its maximum and minimum at the initial time $t = 0$ or on the boundary $x = 0$ or $x = 1$ as a function of two variables defined on the finite strip domain $0 \leq x \leq 1, 0 \leq t \leq T$. This is the well-known *maximum principle* in physics and mathematics.

Since in the maximum principle the boundary value has a significant meaning, we hereafter assume for the boundary condition that

$$a_0(k \, \Delta t) = g_1(k \, \Delta t) \qquad a_n(k \, \Delta t) = g_2(k \, \Delta t) \qquad (8.6.3)$$

Then, if the stability condition (8.6.2) holds, it can be shown that the solution of the finite element scheme (8.5.3) also satisfies the maximum principle.

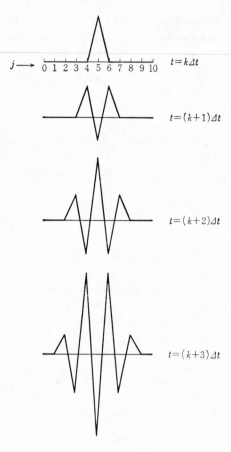

Fig. 8.2 A result with an unstable scheme ($\theta = 0$, $\lambda = \frac{3}{4}$).

Here we take into account the heat source and consider the following scheme, which is obtained by multiplying (8.5.3) by Δt and including an additional inhomogeneous term $\Delta t\, f_j^k$ on the right-hand side:

$$(a_j^{k+1} - a_j^k) + \frac{\sigma\,\Delta t}{h^2} \{\theta(-a_{j-1}^{k+1} + 2a_j^{k+1} - a_{j+1}^{k+1})$$

$$+ (1 - \theta)(-a_{j-1}^k + 2a_j^k - a_{j+1}^k)\} = \Delta t\, f_j^k \qquad (8.6.4)$$

Then this scheme satisfies the maximum principle:

$$\min\,\{g_{\min}^{k+1},\, a_{\min}^k + \Delta t\, f_{\min}^k\} \le a_j^{k+1} \le \max\,\{g_{\max}^{k+1},\, a_{\max}^k + \Delta t\, f_{\max}^k\}$$

$$(8.6.5)$$

The quantities with the subscript max are defined as

$$a_{max}^k = \max_{0 \leq j \leq n} a_j^k \qquad (8.6.6)$$

$$g_{max}^{k+1} = \max \{ g_1((k+1)\,\Delta t), g_2((k+1)\,\Delta t) \} \qquad (8.6.7)$$

$$f_{max}^k = \max_{0 \leq j \leq n} f_j^k \qquad (8.6.8)$$

The quantities with the subscript min are defined by replacing max by min on the right-hand side of (8.6.6) through (8.6.8). The second inequality in (8.6.5) implies that, if there is no heat source, the maximum value of the approximate solution at $t = (k+1)\,\Delta t$ never exceeds the maximum value at $t = k\,\Delta t$ or the boundary value at $t = (k+1)\,\Delta t$. Also, the first inequality in (8.6.5) implies that the same situation holds for the minimum.

The proof of the maximum principle (8.6.5) is exactly the same as that for (8.5.18). We prove here the second inequality in (8.6.5). Since this inequality is trivial when a_j^{k+1} attains its maximum on the boundary at $t = (k+1)\,\Delta t$, we assume that a_j^{k+1} attains its maximum inside the domain, that is, for some $j = m$, where $1 \leq m \leq n-1$. Then, if we put $j = m$ in

$$(1+2\theta\lambda)a_j^{k+1} = \theta\lambda a_{j-1}^{k+1} + \theta\lambda a_{j+1}^{k+1} + (1-\theta)\lambda a_{j-1}^k$$
$$+ (1 - 2(1-\theta)\lambda)a_j^k + (1-\theta)\lambda a_{j+1}^k + \Delta t f_j^k$$
$$(8.6.9)$$

which follows from (8.6.4), the fourth term on the right-hand side is positive under assumption (8.6.2), so that we immediately obtain

$$(1+2\theta\lambda)a_m^{k+1} \leq \theta\lambda a_{m-1}^{k+1} + \theta\lambda a_{m+1}^{k+1} + (1-\theta)\lambda a_{max}^k + (1-2(1-\theta)\lambda)a_{max}^k$$
$$+ (1-\theta)\lambda a_{max}^k + \Delta t f_{max}^k$$
$$\leq 2\theta\lambda a_m^{k+1} + a_{max}^k + \Delta t f_{max}^k$$
$$(8.6.10)$$

that is,

$$a_m^{k+1} \leq a_{max}^k + \Delta t f_{max}^k \qquad (8.6.11)$$

The proof of the first inequality in (8.6.5) for the minimum may be obtained in a similar way.

The stability condition (8.6.2) always holds if $\theta = 1$. This means that a backward lumped mass scheme is unconditionally stable. When $\theta < 1$, on the other hand, if the mesh size h of the space variable is made smaller, the time step Δt must be made much smaller in such a way that it is proportional to h^2.

The scheme described in this section is a simple one in one space dimension. In more general problems the scheme the error satisfies has the same form as the scheme the approximate solution satisfies, with an additional inhomogeneous term r_j^k on the right-hand side as in (8.5.8). The term r_j^k corresponds to the difference between the difference scheme the approximate solution satisfies and the differential equation the exact solution satisfies. As seen above, if r_j^k is appropriately small and the scheme satisfies the maximum principle, the error of the approximate solution is bounded small.

8.7 Stability of the Consistent Mass System

Based on the form of the mass matrix (8.1.13), the jth row of scheme (8.2.7) for the consistent mass system can be written as

$$(\tfrac{1}{6} - \theta\lambda)\, a_{j-1}^{k+1} + (\tfrac{2}{3} + 2\theta\lambda)\, a_j^{k+1} + (\tfrac{1}{6} - \theta\lambda)\, a_{j+1}^{k+1}$$
$$= (\tfrac{1}{6} + (1-\theta)\lambda)\, a_{j-1}^k + (\tfrac{2}{3} - 2(1-\theta)\lambda)\, a_j^k + (\tfrac{1}{6} + (1-\theta)\lambda)\, a_{j+1}^k$$
$$(8.7.1)$$

It is evident from the proof of stability in the previous section that scheme (8.7.1) is stable if the coefficients of the first and third terms on the left-hand side are nonpositive and the coefficient of the second term on the right-hand side in nonnegative. Therefore the stability condition for the consistent mass system is given by

$$\frac{1}{6\theta} \le \frac{\sigma\, \Delta t}{h^2} \qquad (8.7.2)$$

$$\frac{1}{3(1-\theta)} \ge \frac{\sigma\, \Delta t}{h^2} \qquad (8.7.3)$$

The inequality (8.7.3) is more stringent than (8.6.2) and, in addition, (8.7.2) must be satisfied. Thus we conclude that, from the standpoint of stability, the lumped mass system is not only simpler but also better than the consistent mass system.

8.8 Stability Based on Eigenvalues

The stability of a scheme is closely related to the eigenvalues of the coefficient matrix of the system of linear equations. Scheme (8.2.7) can be written formally as

$$a((k+1)\, \Delta t) = (M + \theta\, \Delta t\, K)^{-1}(M - (1-\theta)\, \Delta t\, K)\, a(k\, \Delta t) \qquad (8.8.1)$$

Since the solution is computed iteratively by solving the system of linear equations (8.8.1), the eigenvalues of the matrix

$$(M + \theta \, \Delta t \, K)^{-1}(M - (1 - \theta) \, \Delta t \, K)$$
$$= (I + \theta \, \Delta t \, M^{-1}K)^{-1}(I - (1 - \theta) \, \Delta t \, M^{-1}K) \qquad (8.8.2)$$

on the right-hand side directly affect the stability of the scheme.

Let the initial error be $\epsilon(0)$. Then the computation is actually carried out according to

$$\{a((k + 1) \, \Delta t) + \epsilon((k + 1) \, \Delta t)\}$$
$$= (M + \theta \, \Delta t \, K)^{-1}(M - (1 - \theta) \, \Delta t \, K)\{a(k \, \Delta t) + \epsilon(k \, \Delta t)\} \qquad (8.8.3)$$

which includes the error term $\epsilon(k \, \Delta t)$, so that the error propagates as

$$\epsilon((k + 1) \, \Delta t) = (I + \theta \, \Delta t \, M^{-1}K)^{-1}(I - (1 - \theta) \, \Delta t \, M^{-1}K)\epsilon(k \, \Delta t) \qquad (8.8.4)$$

Let an eigenvalue of matrix (8.8.2) be μ and the corresponding eigenvector be v. Here we trace the change in the component of the initial error corresponding, in particular, to the eigenvector v. To this end we let

$$\epsilon(0) = \epsilon_0 v \qquad (8.8.5)$$

where ϵ_0 is a small number. Then from (8.8.4) the error at time $t = k \, \Delta t$ becomes

$$\epsilon(k \, \Delta t) = \epsilon_0 \mu^k v \qquad (8.8.6)$$

This indicates that, if there is an eigenvalue whose absolute value is larger than 1, the error becomes larger as we repeat the iteration. In other words, to prevent the error from increasing, the absolute value of each of the eigenvalues must be less than or equal to 1. This is an alternative condition required for stability.

8.9 Stability Condition for the Lumped Mass System Based on Eigenvalues

We consider here the eigenvalues of matrix (8.8.2) for the lumped mass system. For this purpose we compute the lth eigenvalue v_l and the corresponding eigenvector $y^{(l)}$ of the matrix $M^{-1}K$ appearing in (8.8.2). Here v_l and $y^{(l)}$ satisfy

$$M^{-1}Ky^{(l)} = v_l y^{(l)} \qquad (8.9.1)$$

which is equivalent to

$$(K - v_l M)y^{(l)} = 0 \qquad (8.9.2)$$

Taking into account the definitions (8.1.14) and (8.3.3) of K and M, we obtain from the jth row of (8.9.2), by multiplying both sides by $-1/h$,

$$-\frac{\sigma}{h^2} y_{j+1}^{(l)} + \left(\nu_l - \frac{2\sigma}{h^2}\right) y_j^{(l)} + \frac{\sigma}{h^2} y_{j-1}^{(l)} = 0 \qquad j = 1,2,\ldots,n-1$$

(8.9.3)

where $y_j^{(l)}$ is the jth element of $\mathbf{y}^{(l)}$. We also require

$$y_0^{(l)} = y_n^{(l)} = 0 \tag{8.9.4}$$

so that (8.9.3) also holds for $j = 1$ and $j = n - 1$.

Then the solution of the difference equation (8.9.3) can be obtained by the substitution

$$y_j^{(l)} = \sin \omega_l j \tag{8.9.5}$$

It is evident that $y_0^{(l)} = 0$ in (8.9.4) holds by itself, but in order to claim that $y_n^{(l)} = \sin \omega_l n = 0$ we need

$$\omega_l = \frac{\pi l}{n} = \pi h l \qquad l = 1,2,\ldots,n-1 \qquad h = \frac{1}{n} \tag{8.9.6}$$

From this requirement we have for the eigenvalues of $M^{-1}K$

$$y_j^{(l)} = \sin \frac{\pi l}{n} j \qquad j = 1,2,\ldots,n-1 \tag{8.9.7}$$

Furthermore, if we substitute (8.9.5) into (8.9.3) we have

$$\left\{\nu_l - \frac{2\sigma}{h^2}(1 - \cos \omega_l)\right\} \sin \omega_l j = 0 \qquad j = 1,2,\ldots,n-1$$

(8.9.8)

In order for these equalities to hold for all $j = 1,2,\ldots,n-1$ with fixed l, ν_l must be

$$\nu_l = \frac{2\sigma}{h^2}\left(1 - \cos \frac{\pi l}{n}\right) = \frac{4\lambda}{\Delta t} \sin^2 \frac{\pi l}{2n} \qquad \lambda = \frac{\sigma \Delta t}{h^2} \tag{8.9.9}$$

Thus we have obtained the lth eigenvalues of $M^{-1}K$.

Since the eigenvalues of $M^{-1}K$ are known, we can immediately write the eigenvalues of (8.8.2). These eigenvalues, which we denote by μ_l, are

$$\mu_l = \frac{1 - (1-\theta)\,\Delta t\,\nu_l}{1 + \theta\,\Delta t\,\nu_l} = \frac{1 - 4(1-\theta)\lambda \sin^2(\pi l/2n)}{1 + 4\theta\lambda \sin^2(\pi l/2n)}$$

(8.9.10)

To prevent the error from being increased by iteration, the absolute value of μ_l must not be larger than one that can be written as

$$-1 \leq \frac{1 - 4(1-\theta)\lambda \sin^2(\pi l/2n)}{1 + 4\theta\lambda \sin^2(\pi l/2n)} \leq 1 \tag{8.9.11}$$

It is evident that the second inequality holds. From the first inequality we have

$$-1 \le 2(2\theta - 1)\lambda \sin^2 \frac{\pi l}{2n} \qquad (8.9.12)$$

which always holds if $\theta \ge \frac{1}{2}$. When $\theta < \frac{1}{2}$, we have

$$\frac{1}{\sin^2 \pi l/2n} \frac{1}{2(1-2\theta)} \ge \lambda \qquad (8.9.13)$$

Since $1/\sin^2(\pi l/2n) \ge 1$, if

$$\frac{1}{2(1-2\theta)} \ge \frac{\sigma \Delta t}{h^2} \qquad (8.9.14)$$

then (8.9.13) holds for any l. Summarizing the above we have the following stability condition such that the error is not increased by iteration:

$$\begin{array}{ll} \text{Unconditionally} & \text{if } \theta \ge \frac{1}{2} \\ \dfrac{1}{2(1-2\theta)} \ge \dfrac{\sigma \Delta t}{h^2} & \text{if } \theta < \frac{1}{2} \end{array} \qquad (8.9.15)$$

Note that the first inequality in (8.9.11) places a particular restriction on eigenvalues of $M^{-1}K$ with large absolute values.

Condition (8.9.15) is much less severe than stability condition (8.6.2) based on the maximum principle. This means that, even if the scheme satisfies (8.9.15), the maximum principle does not always hold. If the scheme satisfies (8.9.15), the error indeed is not increased remarkably by iteration. The absolute value of μ_l, however, approaches 1 if $\lambda = \sigma \Delta t/h^2$ is too large or too small, and then the decay of the error deteriorates. In the case of the heat equation, the scheme should be constructed in such a way that, from a physical point of view, the maximum principle holds, that is, (8.6.2) is satisfied.

In terms of the approximate solution $\hat{u}_n(x,t)$, the stability based on the maximum principle corresponds to the stability condition that $\hat{u}_n$ does not increase abnormally with respect to the L_∞ norm defined by

$$\|\hat{u}_n\|_\infty \equiv \max_{0 \le x \le 1} |\hat{u}_n(x,t)| = \max_{0 \le j \le n} |a_j(t)| \qquad (8.9.16)$$

On the other hand, the stability based on the eigenvalues corresponds to the stability condition that $\hat{u}_n$ does not increase abnormally with respect to the L_2 norm defined by

$$\|\hat{u}_n\|_2 \equiv \left[\int_0^1 |\hat{u}_n(x,t)|^2 \, dx \right]^{1/2} = [a^T M a]^{1/2} \qquad (8.9.17)$$

The former is called the L_∞ *stability*, and the latter is called the L_2 *stability*.

8.10 *Stability Condition for the Consistent Mass System Based on Eigenvalues*

A similar stability condition can be derived for the consistent mass system based on the eigenvalues of the matrix. In this case the difference equation corresponding to (8.9.3) is

$$\left(\frac{1}{6}\nu_l + \frac{\sigma}{h^2}\right) y_{j+1}^{(l)} + \left(\frac{2}{3}\nu_l - \frac{2\sigma}{h^2}\right) y_j^{(l)} + \left(\frac{1}{6}\nu_l + \frac{\sigma}{h^2}\right) y_{j-1}^{(l)} = 0$$

$$j = 1, 2, \ldots, n-1 \tag{8.10.1}$$

and the substitution of (8.9.5) results in a system of equations

$$\left\{\left(\frac{1}{3}\nu_l + \frac{2\sigma}{h^2}\right) \cos \omega_l + \left(\frac{2}{3}\nu_l - \frac{2\sigma}{h^2}\right)\right\} \sin \omega_l j = 0 \quad j = 1, 2, \ldots, n-1 \tag{8.10.2}$$

which determines ω_l. From this we see that the νth eigenvalue of $M^{-1}K$ is

$$\nu_l = \frac{6\sigma}{h^2} \frac{1 - \cos(\pi l/n)}{2 + \cos(\pi l/n)} \tag{8.10.3}$$

Substituting this eigenvalue into $M^{-1}K$ in (8.8.2) and imposing the condition that its absolute value is less than 1, we have a stability condition

$$\begin{array}{cc} \text{Unconditionally} & \text{if } \theta \geq \frac{1}{2} \\ & \\ \dfrac{1}{6(1 - 2\theta)} \geq \dfrac{\sigma \Delta t}{h^2} & \text{if } \theta < \frac{1}{2} \end{array} \tag{8.10.4}$$

which is more severe than (8.9.15) for the lumped mass system. This also includes the condition (8.7.2) and (8.7.3) based on the maximum principle.

As seen from the name of the stiffness matrix for K, $M^{-1}K$ can be regarded as a quantity that corresponds to the stiffness of the physical system being considered. By comparing the eigenvalues (8.10.3) and (8.9.9) we see that the eigenvalue of $M^{-1}K$ is decreased by lumping of the mass, which means that the structure of the physical system becomes less stiff. In particular, a decrease in the largest absolute eigenvalue of $M^{-1}K$ results in a decrease in the eigenvalue of the matrix (8.8.2), which directly affects the stability. Then the oscillatory structure of the solution of the approximate equation (8.2.7) at each time step is modified; hence the system becomes more stable.

9

The Heat Equation in Two Space Dimensions

9.1 Subdivision of a Two-Dimensional Domain and the Barycentric Region

From the derivation of the FEM for the heat equation in one space dimension and that for stationary problems in two space dimensions we can easily see how to apply this procedure to the heat equation in two space dimensions.

We assume here that the shape of the boundary ∂G of the domain G being considered is a convex polygon and consider the following heat equation there:

$$\frac{\partial u(x, y, t)}{\partial t} = \sigma \left\{ \frac{\partial^2 u(x, y, t)}{\partial x^2} + \frac{\partial^2 u(x, y, t)}{\partial y^2} \right\} \tag{9.1.1}$$

$$u(x, y, t) = g(x, y, t) \qquad \text{on } G \tag{9.1.2}$$

$$u(x, y, 0) = u_0(x, y) \tag{9.1.3}$$

where σ is assumed constant. We also assume a compatibility condition at $t = 0$.

$$u_0(x, y) = g(x, y, 0) \tag{9.1.4}$$

In order to apply the FEM to this problem we divide the domain G into triangular subdomains as in the case of stationary problems, that is, the elliptic boundary value problems discussed in Chap. 4, denote each

113

node by P_k, and construct the basis function $\hat{\phi}_k(x,y)$ consisting of piecewise linear polynomials that satisfy

$$\hat{\phi}_k(x_j, y_j) = \begin{cases} 1 & j = k \\ 0 & j \neq k \end{cases} \tag{9.1.5}$$

The boundary condition (9.1.2) is approximated by linear interpolation based on the values of $g(x,y,t)$ at the nodal points on the boundary. Thus we expand the approximate solution $u_n(x,y,t)$ to $u(x,y,t)$ as

$$\hat{u}_n(x,y,t) = \sum_{j=1}^{n} a_j(t)\hat{\phi}_j(x,y) + \sum_{j=n+1}^{n+\nu} b_j(t)\hat{\phi}_j(x,y) \tag{9.1.6}$$

$$b_j(t) = g(x_j, y_j, t) \qquad j = n+1, n+2, \ldots, n+\nu \tag{9.1.7}$$

The nodes from 1 to n correspond to the internal nodes, and those from $n+1$ to $n+\nu$ correspond to the boundary nodes. The basis functions corresponding to the boundary nodes are defined as identically zero outside the domain G.

In problems in one space dimension we observed the merit of lumping of the mass, so in this section we employ lumping of the mass from the beginning. Although there are several methods of lumping, we use here a procedure based on the barycenter. Let the barycenter of the triangular element τ whose nodes are P_1, P_2, and P_3 be G_τ, and let the midpoints of P_1P_2, P_2P_3, and P_3P_1 be P_3', P_1', and P_2', respectively, as shown in Fig. 9.1. We divide τ into three subdomains by three line segments $G_\tau P_3'$, $G_\tau P_1'$, and $G_\tau P_2'$ and assign each subdomain to the corresponding node. For example, the quadrilateral $P_1 P_3' G_\tau P_2'$ is assigned to the node P_1.

After making this assignment for all the nodes, we combine the subdomains assigned to a particular node into a region. This region is called the barycentric region corresponding to the node. Then we define a piecewise constant function

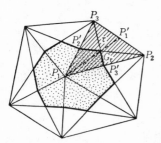

Fig. 9.1 Barycentric region corresponding to the node P_1 (dotted region).

$$\bar{\varphi}_j(x,y) \tag{9.1.8}$$

whose value is 1 in the barycentric region corresponding to the jth node and is 0 otherwise. This can be regarded as a lumping version of $\hat{\phi}_j(x,y)$. It is evident that the set of functions $\{\bar{\phi}_j\}$ satisfies a simple orthogonal relation

$$\iint_G \bar{\varphi}_j(x,y)\bar{\varphi}_k(x,y) \, dx \, dy = 0 \qquad j \neq k \tag{9.1.9}$$

9.2 Application of the Finite Element Method

Now we apply the FEM to the heat equation. We substitute $\hat{u}_n(x,y,t)$ in (9.1.6) for u in (9.1.1), multiply both sides by $\hat{\phi}_k(x,y)$, $k=1,2,\ldots,n$, and integrate by parts using (4.1.9). Then we have

$$
\begin{aligned}
&\sum_{j=1}^n \frac{da_j}{dt} \iint_G \hat{\varphi}_j\hat{\varphi}_k \, dx \, dy + \sum_{j=n+1}^{n+\nu} \frac{db_j}{dt} \iint_G \hat{\varphi}_j\hat{\varphi}_k \, dx \, dy \\
&+ \sigma \sum_{j=1}^n a_j \iint_G \left(\frac{\partial\hat{\varphi}_j}{\partial x} \frac{\partial\hat{\varphi}_k}{\partial x} + \frac{\partial\hat{\varphi}_j}{\partial y} \frac{\partial\hat{\varphi}_k}{\partial y} \right) dx \, dy \\
&+ \sigma \sum_{j=n+1}^{n+\nu} b_j \iint_G \left(\frac{\partial\hat{\varphi}_j}{\partial x} \frac{\partial\hat{\varphi}_k}{\partial x} + \frac{\partial\hat{\varphi}_j}{\partial y} \frac{\partial\hat{\varphi}_k}{\partial y} \right) dx \, dy \\
&- \sigma \sum_{j=1}^n a_j \int_{\partial G} \frac{\partial\hat{\varphi}_j}{\partial n} \hat{\varphi}_k \, d\sigma - \sigma \sum_{j=n+1}^{n+\nu} b_j \int_{\partial G} \frac{\partial\hat{\varphi}_j}{\partial n} \hat{\varphi}_k \, d\sigma = 0
\end{aligned}
\tag{9.2.1}
$$

Since each function $\hat{\phi}_k(x,y)$ multiplied by both sides is a basis function corresponding to an internal node, the boundary integrals in (9.2.1) vanish.

We now apply lumping of the mass. We carry out the replacement

$$\iint_G \hat{\varphi}_j\hat{\varphi}_k \, dx \, dy \rightarrow \iint_G \bar{\varphi}_j\bar{\varphi}_k \, dx \, dy \tag{9.2.2}$$

on the left-hand side of (9.2.1). The second integral of (9.2.1) vanishes because k and j in the integral never coincide with each other. From this we see that (9.2.1) can be written in matrix form as

$$M \frac{d\boldsymbol{a}}{dt} + K\boldsymbol{a} = -L\boldsymbol{b} \tag{9.2.3}$$

where M is an $n \times n$ mass matrix whose jk entry is

$$M_{jk} = \iint_G \bar{\varphi}_j\bar{\varphi}_k \, dx \, dy \tag{9.2.4}$$

and is diagonal because of lumping of the mass. Here K and L are the $n \times n$ and $n \times \nu$ stiffness matrices, respectively, whose jk entries are given by

$$K_{jk}, L_{jk} = \sigma \iint_G \left(\frac{\partial \hat{\varphi}_j}{\partial x} \frac{\partial \hat{\varphi}_k}{\partial x} + \frac{\partial \hat{\varphi}_j}{\partial y} \frac{\partial \hat{\varphi}_k}{\partial y} \right) dx \, dy \qquad (9.2.5)$$

The jk entry of L does not vanish only when j or k corresponds to the boundary node or to the node adjacent to the boundary. In order to compute these entries we use element matrices as in the case of the stationary problem. The vectors a and b are defined as

$$a = a(t) = \begin{pmatrix} a_1(t) \\ a_2(t) \\ \vdots \\ a_n(t) \end{pmatrix} \qquad (9.2.6)$$

$$b = b(t) = \begin{pmatrix} b_{n+1}(t) \\ b_{n+2}(t) \\ \vdots \\ b_{n+\nu}(t) \end{pmatrix} \qquad (9.2.7)$$

As seen from (9.1.7) b is a known vector.

The stiffness matrices K and L satisfy an important equality. Consider a constant function $w(x,y)$ whose value is identically unity in the internal domain G and can be expressed as

$$w(x,y) = 1 = \sum_{j=1}^n \hat{\varphi}_j(x,y) + \sum_{j=n+1}^{n+\nu} \hat{\varphi}_j(x,y) \qquad (9.2.8)$$

By differentiating we have

$$\frac{\partial w}{\partial x} = \frac{\partial w}{\partial y} = 0 \qquad (9.2.9)$$

Therefore we obtain

$$0 = \sigma \iint_G \left(\frac{\partial \hat{\varphi}_i}{\partial x} \frac{\partial w}{\partial x} + \frac{\partial \hat{\varphi}_i}{\partial y} \frac{\partial w}{\partial y} \right) dx \, dy$$

$$= \sum_{j=1}^n K_{ij} + \sum_{j=n+1}^{n+\nu} L_{ij} \qquad i = 1, 2, \dots, n \qquad (9.2.10)$$

This implies that the sum of all the entries of an arbitrary row of the stiffness matrix, including the entries corresponding to the boundary nodes, vanishes.

We discretize the time t with a step size Δt, as in the one-dimensional case, and replace the time derivative by the time difference in the system of ordinary differential equations (9.2.3). We also introduce the parameter θ and mix the forward difference with $\theta = 0$ and the backward difference with $\theta = 1$:

$$M \frac{a((k + 1)\ \Delta t) - a(k\ \Delta t)}{\Delta t} + K\{\theta a((k + 1)\ \Delta t) + (1 - \theta)a(k\ \Delta t)\}$$

$$= -L\{\theta b((k + 1)\ \Delta t) + (1 - \theta)b(k\ \Delta t)\} \qquad 0 \le \theta \le 1 \qquad (9.2.11)$$

After multiplying both sides by Δt and performing a slight manipulation, we have

$$(M + \theta\ \Delta t\ K)a((k + 1)\ \Delta t) = (M - (1 - \theta)\ \Delta t\ K)a(k\ \Delta t)$$

$$-\theta\ \Delta t\ Lb((k + 1)\ \Delta t) - (1 - \theta)\ \Delta t\ Lb(k\ \Delta t) \qquad (9.2.12)$$

If we assume, for example, as an initial value

$$a(0) = u_0 = \begin{pmatrix} u_0(x_1, y_1) \\ u_0(x_2, y_2) \\ \vdots \\ u_0(x_n, y_n) \end{pmatrix} \qquad (9.2.13)$$

and solve (9.2.12) repeatedly for $a(t)$ for $t = \Delta t, 2\Delta t, \ldots$, we obtain a finite element solution $\hat{u}_n(x, y, t)$ in the form of (9.1.6). Since M is a diagonal matrix, $a((k + 1)\ \Delta t)$ is obtained from (9.2.12) by simple division when $\sigma = 0$.

9.3 The Maximum Principle for a Lumped Mass System and Triangulation of the Acute Type

As will be shown below, in order to obtain a stable solution (9.2.12) must satisfy the following *maximum principle* as in the one-dimensional case:

$$\min\ \{a_{\min}^k, b_{\min}^k, b_{\min}^{k+1}\} \le a_j((k + 1)\ \Delta t)$$

$$\le \max\ \{a_{\max}^k, b_{\max}^k, b_{\max}^{k+1}\} \qquad j = 1, 2, \ldots, n \qquad (9.3.1)$$

where

$$a_{\min}^k = \min_{1 \le j \le n} a_j(k\ \Delta t) \qquad (9.3.2)$$

$$a_{\max}^k = \max_{1 \le j \le n} a_j(k\ \Delta t) \qquad (9.3.3)$$

$$b_{\min}^k = \min_{n+1 \le j \le n+\nu} b_j(k\ \Delta t) \qquad (9.3.4)$$

$$b_{\max}^k = \max_{n+1 \le j \le n+\nu} b_j(k\ \Delta t) \qquad (9.3.5)$$

Before proving (9.3.1) we make an important assumption. We assume that the triangulation of the domain G is of the acute type. This means that no triangular element has an obtuse angle. If the triangulation is of the acute type, we have for the stiffness matrix:

$$K_{ij} \le 0 \qquad L_{ij} \le 0 \qquad i \ne j \qquad (9.3.6)$$

from (5.7.2).

In order to prove the maximum principle (9.3.1) we write the ith row of (9.2.12) explicitly:

$$\sum_{j=1}^{n} (M_{ij} + \theta \, \Delta t \, K_{ij}) \, a_j((k+1) \, \Delta t)$$

$$= \sum_{j=1}^{n} (M_{ij} - (1-\theta) \, \Delta t \, K_{ij}) a_j(k \, \Delta t) - \theta \, \Delta t \sum_{j=n+1}^{n+v} L_{ij} b_j((k+1) \, \Delta t)$$

$$- (1-\theta) \, \Delta t \sum_{j=n+1}^{n+v} L_{ij} b_j(k \, \Delta t) \qquad i = 1, 2, \ldots, n \qquad (9.3.7)$$

Note that the mass matrix M is diagonal; that is, $M_{ij} = 0$ for $i \ne j$ since we have employed lumping of the mass. We move the terms corresponding to the off-diagonal entries on the left-hand side to the right-hand side. Then (9.3.7) becomes

$$(M_{ii} + \theta \, \Delta t \, K_{ii}) a_i((k+1) \, \Delta t) = (M_{ii} - (1-\theta) \, \Delta t \, K_{ii}) a_i(k \, \Delta t)$$

$$- (1-\theta) \, \Delta t \sum_{\substack{j=1 \\ j \ne i}}^{n} K_{ij} a_j(k \, \Delta t) - \theta \, \Delta t \sum_{\substack{j=1 \\ j \ne i}}^{n} K_{ij} a_j((k+1) \, \Delta t)$$

$$- \theta \, \Delta t \sum_{j=n+1}^{n+v} L_{ij} b_j((k+1) \, \Delta t) - (1-\theta) \, \Delta t \sum_{j=n+1}^{n+v} L_{ij} b_j(k \, \Delta t)$$

$$(9.3.8)$$

By analogy with the one-dimensional case we see that (9.3.8) is stable if all the coefficients on both sides are nonnegative. The coefficient whose sign is still indefinite is that of the first term on the right-hand side. It is evident that each of the other coefficients is positive or zero if the triangulation is of the acute type. Thus, leaving a detailed discussion to the next section, we assume at this stage that

$$M_{ii} - (1-\theta) \, \Delta t \, K_{ii} \ge 0 \qquad i = 1, 2, \ldots, n \qquad (9.3.9)$$

We prove here the second inequality of the maximum principle (9.3.1). Since it is evident that the equality sign holds for the boundary nodes, we assume that $a_i((k+1) \, \Delta t)$ attains the maximum a_m at a certain internal node $i = m$, $1 \le m \le n$. We denote the maximum value of $a_j(k \, \Delta t)$ and

$b_j(k \, \Delta t)$ at the previous time step by $a_{m'}$, that is,

$$a_j((k + 1) \, \Delta t), b_j((k + 1) \, \Delta t) \le a_m$$

$$a_j(k \, \Delta t), b_j(k \, \Delta t) \le a_{m'} \tag{9.3.10}$$

Then, based on (9.3.9) and assumption (9.3.6) for triangulation of the acute type, we obtain the following inequality by substituting $i = m$ in (9.3.8):

$$M_{mm} a_m + \theta \, \Delta t \, K_{mm} a_m \le M_{mm} a_{m'} - (1 - \theta) \, \Delta t \left(\sum_{j=1}^{n} K_{mj} + \sum_{j=n+1}^{n+v} L_{mj} \right) a_{m'}$$

$$- \theta \, \Delta t \left(\sum_{\substack{j=1 \\ j \ne m}}^{n} K_{mj} + \sum_{j=n+1}^{n+v} L_{mj} \right) a_m \tag{9.3.11}$$

Moving the second term on the left-hand side to the right-hand side and using (9.2.10), we obtain $a_m \le a_{m'}$. Thus the second inequality of (9.3.1) is proved under the assumption of triangulation of the acute type and (9.3.9). The proof of the first term with respect to the minimum is quite similar to that shown above.

9.4 A Sufficient Condition for Stability

Now we discuss assumption (9.3.9). Let the element mass matrix and the element stiffness matrix for element τ be m^τ and k^τ, respectively. If

$$m_{ii}^\tau - (1 - \theta) \, \Delta t \, k_{ii}^\tau \ge 0 \tag{9.4.1}$$

holds for every τ, (9.3.9) evidently holds. So we make a detailed analysis of (9.4.1). Since m_{ii}^τ is equal to the area of the intersection of τ and the barycentric region corresponding to node i, we have

$$m_{ii}^\tau = \tfrac{1}{3} |S| \tag{9.4.2}$$

and from (5.7.2) we also have

$$k_{ii}^\tau = \frac{\sigma}{4|S|} |q_i|^2 \tag{9.4.3}$$

where $|S|$ is equal to the area of τ and $|q_i|$ is the distance from the opposite side to node i. Substitution of (9.4.2) and (9.4.3) into (9.4.1) results in

$$\frac{1}{3} |S| - (1 - \theta) \, \Delta t \, \frac{\sigma}{4|S|} |q_i|^2 = \frac{|S|}{3\kappa_i^2} \{ \kappa_i^2 - 3\sigma(1 - \theta) \, \Delta t \} \tag{9.4.4}$$

where κ_i is the length of a perpendicular line from the ith node to the opposite side. If we write

$$\kappa_{min} = \min \kappa_i \tag{9.4.5}$$

we conclude that under the assumption that

$$\kappa_{min}^2 \geq 3\sigma(1 - \theta) \, \Delta t \tag{9.4.6}$$

(9.4.1) always holds. This corresponds to condition (8.6.2) in the one-dimensional case and is a constraint on the time step Δt, which depends on triangulation of the domain. Note that (9.4.6) always holds if $\theta = 1$.

To summarize, it has become apparent that, if the triangulation is of the acute type and the time step size Δt satisfies (9.4.6), then the maximum principle holds for scheme (9.2.12) and is stable.

9.5   *The Maximum Principle for a Consistent Mass System*

In the case of one space dimension we observed that the lumped mass system was more stable than the consistent mass system. This situation also holds in the case of two space dimensions. As we have seen so far, the maximum principle holds if the coefficient of every term of (9.3.8) is positive or zero. Therefore, if the inequalities

$$M_{ii} - (1 - \theta) \, \Delta t \, K_{ii} \geq 0 \tag{9.5.1}$$

$$M_{ij} + \theta \, \Delta t \, K_{ij} \leq 0 \qquad i \neq j \tag{9.5.2}$$

hold in (9.3.7), the maximum principle holds for the consistent mass system also.

We also make a detailed analysis of these inequalities for each element τ. First, from (5.7.1) and (5.7.2), if

$$\kappa_{min}^2 \geq 6\sigma(1 - \theta) \, \Delta t \tag{9.5.3}$$

then (9.5.1) holds, as in the case of the lumped mass system. Next, (9.5.2) holds if

$$m_{ij}^\tau + \theta \, \Delta t \, k_{ij}^\tau = \frac{1}{12} S - \theta \, \Delta t \, \frac{\sigma}{4S} \, |q_i| \, |q_j| \cos \theta_k$$
$$\leq S \left[\frac{1}{12} - \sigma\theta \, \Delta t \, \frac{\cos \theta_{max}}{\kappa_{max}^2} \right] \leq 0 \tag{9.5.4}$$

that is, if

$$\kappa_{max}^2 \leq 12\sigma\theta \, \Delta t \cos \theta_{max} \tag{9.5.5}$$

where

$$\kappa_{max} = \max \kappa_i \tag{9.5.6}$$

$$\theta_{\max} = \max\ \theta_i \qquad\qquad (9.5.7)$$

From (9.5.5) we see that, in the consistent mass system, the stability may fail to hold if the triangulation includes a triangle with a right angle, let alone one with an obtuse angle. Condition (9.5.3) is more severe than (9.4.6) and, in addition, (9.5.5) is required. Thus the consistent mass system is also generally less stable than the lumped mass system in two-dimensional cases.

10

The Wave Equation

10.1 *The Finite Element Method for the Wave Equation*

The finite element formulation of the wave problem or the oscillation problem is in principle similar to that of the heat conduction problem from a technical point of view. For simplicity we consider the following wave problem in one space dimension:

$$\frac{\partial^2 u}{\partial t^2} = c^2 \frac{\partial^2 u}{\partial x^2} \tag{10.1.1}$$

$$u(0,t) = u(1,t) = 0 \tag{10.1.2}$$

$$u(x,0) = \psi_1(x) \tag{10.1.3}$$

$$\frac{\partial u}{\partial t}(x,0) = \psi_2(x) \tag{10.1.4}$$

where c is assumed to be a constant and $\psi_1(x)$ and $\psi_2(x)$ are the given initial functions. We divide the interval $(0,1)$ into n equal subintervals, choose $n+1$ points $x_j = jh$, $j = 0,1,\ldots,n$, for the nodes, and write an approximate solution $\hat{u}_n(x,t)$ in the same form as (8.1.5):

$$\hat{u}_n(x,t) = \sum_{j=1}^{n-1} a_j(t)\hat{\varphi}_j(x) \tag{10.1.5}$$

Substituting this expression for u in (10.1.1) and integrating by parts

122

based on the boundary condition (10.1.2), we obtain the following system of ordinary differential equations of second order:

$$M \frac{d^2 a}{dt^2} + Ka = 0 \tag{10.1.6}$$

where M and K are defined by (8.1.3) and (8.1.14), respectively, where σ is replaced by c^2, and a is a vector given by (8.1.8).

10.2 *Mode Superposition*

We present here two methods for solving the system of equations (10.1.6). The first is *mode superposition* in which the initial vector $a(0)$ is decomposed into eigen components or eigen modes of the matrix $M^{-1}K$ and the solution is expressed in terms of their superposition.

We consider the eigenvalue problem

$$Ky = \nu My \tag{10.2.1}$$

which corresponds to (10.1.6). Let the lth eigenvalue be ν_l and the corresponding eigenvector be $y^{(l)}$. Then, as is well known, the solution of (10.1.6) can be written as

$$a(t) = \sum_{l=1}^{n-1} (c_l \cos \sqrt{\nu_l} t + d_l \sin \sqrt{\omega_l} t) y^{(l)} \tag{10.2.2}$$

The coefficients c_l and d_l can be determined so that the solution fits the initial condition:

$$a_j(0) = \sum_{l=1}^{n-1} c_l y_j^{(l)} = \psi_1(x_j) \qquad j = 1, 2, \ldots, n-1 \tag{10.2.3}$$

$$\frac{da_j}{dt}(0) = \sum_{l=1}^{n-1} d_l \sqrt{\nu_l} y_j^{(l)} = \psi_2(x_j) \qquad j = 1, 2, \ldots, n-1 \tag{10.2.4}$$

where $y_j^{(l)}$ is the jth entry of $y^{(l)}$. Note that the eigenvalue problem (10.2.1) is equivalent to (8.9.1), hence we already have the eigenvalues ν_l in hand as (8.10.3). They are all positive.

In problems involving two or three space dimensions, the order of the matrices M and K is usually fairly high. In such cases it is not always necessary to compute all the eigenvalues and eigenvectors of problem (10.2.1) exactly. It is often sufficient in practice to compute several of the smallest or largest eigenvalues and the corresponding eigenvectors and superpose their modes.

If we use the lumped mass system, the mass matrix becomes diagonal and the eigenvalue problem is of a standard form

$$Ky = vhy \qquad (10.2.5)$$

10.3 *Newmark's β-Scheme*

The second method for solving the system of ordinary differential equations (10.1.6) is to integrate it directly with respect to time t with a fixed interval Δt. A direct method is to approximate the second derivative of (10.1.6) by the second difference. However, when we also want to compute the first-order derivative or to solve a more general system of equations having the first-order derivative

$$M \frac{d^2a}{dt^2} + C \frac{da}{dt} + Ka = f \qquad (10.3.1)$$

it is desirable to construct a scheme by which not only a but also its first-order derivative can be computed. Thus we employ the following approximation based on the Taylor expansion for computing the solution at $t + \Delta t$ from the values at t:

$$\dot{a}(t + \Delta t) = \dot{a}(t) + \Delta t \frac{\ddot{a}(t + \Delta t) + \ddot{a}(t)}{2} \qquad (10.3.2)$$

$$a(t + \Delta t) = a(t) + \Delta t \, \dot{a}(t) + \frac{1}{2!} \Delta t^2 \, \ddot{a}(t) + \frac{1}{3!} \Delta t^3 \frac{\ddot{a}(t + \Delta t) - \ddot{a}(t)}{\Delta t}$$

where

$$\dot{a}(t) = \frac{da}{dt} \qquad (10.3.4)$$

$$\ddot{a}(t) = \frac{d^2a}{dt^2} \qquad (10.3.5)$$

It is evident that (10.3.2) and (10.3.3) hold only approximately for general a. However, if we assume that $\ddot{a}(t)$ is linear with respect to t, then from

$$\frac{d^3a}{dt^3} = \frac{\ddot{a}(t + \Delta t) - \ddot{a}(t)}{\Delta t} \qquad (10.3.6)$$

$$\frac{d^ka}{dt^k} = 0 \qquad k \geq 4 \qquad (10.3.7)$$

(10.3.3) holds exactly. In this case we see that, when (10.3.6) is substituted into the derivative of (10.3.3), (10.3.2) also holds exactly. The approximation based on (10.3.2) and (10.3.3) is called the *linear accelera-tion method*.

Although linear acceleration is easy to understand, as shown above, it is not the best way to use (10.3.3), as it is based on the stability, as will be shown in the next section. Thus, in order to improve the stability, we use the following scheme instead of (10.3.3), replacing the coefficient 1/3! of the fourth term on the right-hand side by β:

$$a(t + \Delta t) = a(t) + \Delta t\, \dot{a}(t) + \frac{1}{2!}\, \Delta t^2\, \ddot{a}(t) + \beta\, \Delta t^3 \frac{\ddot{a}(t + \Delta t) - \ddot{a}(t)}{\Delta t}$$

$$= a(t) + \Delta t\, \dot{a}(t) + \beta\, \Delta t^2\, \ddot{a}(t + \Delta t) + \left(\frac{1}{2} - \beta\right) \Delta t^2\, \ddot{a}(t)$$

$$(10.3.8)$$

A scheme based on this approximation can be constructed as follows. First we solve (10.3.8) for $\ddot{a}(t + \Delta t)$, obtaining

$$\ddot{a}(t + \Delta t) = \frac{1}{\beta\, \Delta t^2}\, \{a(t + \Delta t) - a(t)\} - \frac{1}{\beta \Delta t}\, \dot{a}(t) - \left(\frac{1}{2\beta} - 1\right)\ddot{a}(t)$$

$$(10.3.9)$$

Next, substituting this expression into (10.1.6), where time t is replaced by $t + \Delta t$, we have

$$\{M + \beta\, \Delta t^2\, K\}a(t + \Delta t) = Ma(t) + \Delta t\, M\dot{a}(t) + (\tfrac{1}{2} - \beta)\, \Delta t^2\, M\ddot{a}(t)$$

$$(10.3.10)$$

Summarizing, we have the following scheme. First we set the initial values as follows from (10.1.3), (10.1.4), and (10.1.5), respectively,

$$a(0) = \begin{pmatrix} \psi_1(x_1) \\ \psi_1(x_2) \\ \vdots \\ \psi_1(x_{n-1}) \end{pmatrix} \qquad (10.3.11)$$

$$\dot{a}(0) = \begin{pmatrix} \psi_2(x_1) \\ \psi_2(x_2) \\ \vdots \\ \psi_2(x_{n-1}) \end{pmatrix} \qquad (10.3.12)$$

$$\ddot{a}(0) = -M^{-1}Ka(0) \qquad (10.3.13)$$

Next, assuming that $a(t)$, $\dot{a}(t)$, and $\ddot{a}(t)$ at time t are known, we solve (10.3.10) for $a(t + \Delta t)$. Then, using $a(t + \Delta t)$, we compute $\ddot{a}(t + \Delta t)$ from (10.3.9) and $\dot{a}(t + \Delta t)$ from (10.3.2). We repeat this procedure for

$t = k\,\Delta t,\, k = 0,1,2,\dots$. The scheme presented here is called *Newmark's β-scheme* and can be easily generalized for (10.3.1).

10.4 Stability of Newmark's β-Scheme

In this section we discuss the stability of Newmark's β-scheme. Since in this scheme the solution is computed under the assumption that (10.1.6), (10.3.8), and (10.3.2) hold simultaneously, we eliminate $\ddot{a}$ from (10.3.8) and (10.3.2) using (10.1.6). First we eliminate $\ddot{a}(t + \Delta t)$ and $\ddot{a}(t)$ from (10.3.8) using (10.1.6), obtaining

$$\{I + \beta\,\Delta t^2\,M^{-1}K\}a(t+\Delta t) = \{I - (\tfrac{1}{2}-\beta)\,\Delta t^2\,M^{-1}K\}a(t) + \Delta t\,\dot{a}(t)$$
(10.4.1)

Similarly eliminating $\ddot{a}(t + \Delta t)$ and $\ddot{a}(t)$ from (10.3.2) using (10.1.6), we have

$$\frac{\Delta t}{2}\,M^{-1}Ka(t+\Delta t) + \dot{a}(t+\Delta t) = -\frac{\Delta t}{2}\,M^{-1}Ka(t) + \dot{a}(t)$$
(10.4.2)

Let the errors of $a(t)$ and $\dot{a}(t)$ be $\epsilon(t)$ and $\delta(t)$, respectively. Then, from (10.4.1) and (10.4.2) we see that these errors satisfy

$$\{I + \beta\,\Delta t^2 M^{-1}K\}\epsilon(t+\Delta t) = \left\{I - \left(\frac{1}{2} - \beta\right)\Delta t^2 M^{-1}K\right\}\epsilon(t) + \Delta t\,\delta(t)$$
(10.4.3)

$$\frac{\Delta t}{2}\,M^{-1K}\epsilon(t+\Delta t) + \delta(t+\Delta t) = -\frac{\Delta t}{2}\,M^{-1}K\,\epsilon(t) + \delta(t)$$
(10.4.4)

We denote, as in Sec. 8.9, the *l*th eigenvalue and the corresponding eigenvector of the eigenvalue problem (10.2.1) by ν_l and $y^{(l)}$, respectively. Let the components of the initial errors corresponding to this eigenvector be

$$\epsilon(0) = \epsilon_0 y^{(l)}$$
(10.4.5)

$$\delta(0) = \delta_0 y^{(l)}$$
(10.4.6)

where ϵ_0 and δ_0 are small numbers. These initial errors change according to (10.4.3) and (10.4.4), and at $t = k\,\Delta t$ they become

$$\epsilon(k\,\Delta t) = \epsilon_k y^{(l)}$$
(10.4.7)

$$\delta(k\,\Delta t) = \delta_k y^{(l)}$$
(10.4.8)

Substituting these relations into (10.4.3) and (10.4.4), we see that the errors change according to

$$
\begin{bmatrix} 1 + \beta \, \Delta t^2 \, \nu_l & 0 \\[2mm] \dfrac{\Delta t}{2} \, \nu_l & 1 \end{bmatrix} \begin{bmatrix} \epsilon_{k+1} \\[2mm] \delta_{k+1} \end{bmatrix} = \begin{bmatrix} 1 - \left(\dfrac{1}{2} - \beta\right) \Delta t^2 \, \nu_l & \Delta t \\[2mm] -\dfrac{\Delta t}{2} \, \nu_l & 1 \end{bmatrix} \begin{bmatrix} \epsilon_k \\[2mm] \delta_k \end{bmatrix}
$$

$$(10.4.9)$$

To prevent the error from becoming large with increasing k, that is, to be certain that Newmark's β-scheme is stable, the absolute eigenvalues of the matrix

$$
\begin{bmatrix} 1 + \beta \, \Delta t^2 \, \nu_l & 0 \\[2mm] \dfrac{\Delta t}{2} \, \nu_l & 1 \end{bmatrix}^{-1} \begin{bmatrix} 1 - \left(\dfrac{1}{2} - \beta\right) \Delta t^2 \, \nu_l & \Delta t \\[2mm] -\dfrac{\Delta t}{2} \, \nu_l & 1 \end{bmatrix}
$$

$$(10.4.10)$$

which corresponds to (10.4.9), must not be larger than 1. The eigenvalue of this matrix, which we denote by μ, satisfies

$$
\begin{vmatrix} \mu\{1 + \beta \, \Delta t^2 \, \nu_l\} - 1 + \left(\dfrac{1}{2} - \beta\right) \Delta t^2 \, \nu_l & -\Delta t \\[2mm] \mu \dfrac{\Delta t}{2} \, \nu_l + \dfrac{\Delta t}{2} \, \nu_l & \mu = 1 \end{vmatrix} = 0 \qquad (10.4.11)
$$

that is,

$$
\mu^2 - \frac{2\{1 - (\tfrac{1}{2} - \beta) \, \Delta t^2 \, \nu_l\}}{1 + \beta \, \Delta t^2 \, \nu_l} \mu + 1 = 0 \qquad (10.4.12)
$$

Since the eigenvalue ν_l is positive from (8.9.9) or (8.10.3), the coefficient of the first-order term in the quadratic equation (10.4.12) is real. On the other hand, the product of the roots of this equation is 1, so that they cannot be real distinct roots because the absolute value of either root must be larger than 1. Therefore the discriminant of (10.4.12) must be negative or zero. This provides a necessary condition for stability:

$$
\left\{ \frac{1 - (\tfrac{1}{2} - \beta) \, \Delta t^2 \, \nu_l}{1 + \beta \, \Delta t^2 \, \nu_l} \right\}^2 - 1 \leq 0 \qquad (10.4.13)
$$

or

$$
\tfrac{1}{4}(1 - 4\beta) \, \Delta t^2 \, \nu_l \leq 1 \qquad (10.4.14)
$$

If this inequality holds, both of the absolute eigenvalues are 1, and, because of the nature of the wave equation, the absolute value of the error does not increase although it does not diminish.

Consider the case of the lumped mass system. Since $\nu_l \leq 4c^2/h^2$ from (8.9.9), if

$$(1 - 4\beta)\,\frac{c^2\,\Delta t^2}{h^2} \leq 1 \qquad (10.4.15)$$

then (10.4.14) always holds. From this we obtain the stability condition of Newmark's β-scheme for the lumped mass system:

$$\text{Unconditionally} \qquad \text{if } \beta \geq \tfrac{1}{4}$$
$$\frac{c^2\,\Delta t^2}{h^2} \leq \frac{1}{1 - 4\beta} \qquad \text{if } \beta < \tfrac{1}{4} \qquad (10.4.16)$$

As stated in the previous section, the linear acceleration scheme in which β is fixed at 1/3! from the beginning may become unstable for some Δt.

11

A Diffusion Problem With a
Convection Term

11.1 *A One-Dimensional Diffusion Problem With a Convection Term*

One of the most important and most difficult problems to solve is a diffusion problem with a convection term, for example, one involving the diffusion of a pollutant in a flow. In this section we take a one-dimensional model problem as follows and consider some characteristics peculiar to equations with a convection term:

$$\frac{\partial u}{\partial t} = \sigma \frac{\partial^2 u}{\partial x^2} - b(x) \frac{\partial u}{\partial x} + f(x) \qquad 0 < x < 1 \qquad (11.1.1)$$

$$u(0, t) = u(1, t) = 0 \qquad (11.1.2)$$

$$u(x, 0) = u_0(x) \qquad (11.1.3)$$

The term $-b(x) \, \partial u/\partial x$ on the right-hand side is the *convection term*. Here $b(x)$ describes, for example, a known velocity of the flow of clean water, and (11.1.1) describes the diffusion of a contaminant in the given flow, that is, the change in its density u.

Suppose that $b(x) > 0$ in the neighborhood of a point $x = x_a$. If $\partial u/\partial x$, the gradient of u, for $x = x_a$ at time t is positive as shown in Fig. 11.1, $-b(x) \, \partial u/\partial x$ on the right-hand side of (11.1.1) becomes negative as a whole, and the contribution of this term to $\partial u/\partial t$, that is, the change in u at $x = x_a$ with respect to time, results in a decrease in u. From the standpoint of the motion of u as a whole, this means that $b(x) > 0$ in $-b(x) \, \partial u/\partial x$

129

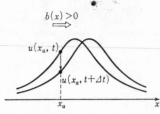

Fig. 11.1 Relation between the sign of $b(x)$ and the direction of the flow.

implies a flow to the right. Thus, in view of the contribution of the term $-b(x)\,\partial u/\partial x$ to $\partial u/\partial t$, we see that $b(x) > 0$ gives a flow toward the right while $b(x) < 0$ gives a flow toward the left. For simplicity we assume that $b(x)$ does not depend on time.

11.2 Application of the Finite Element Method to a One-Dimensional Problem

First we divide the interval $(0, 1)$ into n equal subintervals, construct piecewise linear polynomials $\hat{\phi}_j(x)$ as the basis function, as in (1.3.3), and express an approximate solution of the problem presented in the previous section in terms of $\hat{\phi}_j(x)$:

$$\hat{u}_n(x, t) = \sum_{i=1}^{n-1} a_i(t)\hat{\varphi}_i(x) \tag{11.2.1}$$

Next we substitute this expression for u in (11.1.1), multiply by $\hat{\phi}_j$, and integrate over $(0, 1)$. At this stage of integration we approximate $b(x)$ by its value at the jth node as

$$b(x) \doteq b(x_j) = b_j \qquad x_{j-1} \leq x < x_{j+1} \tag{11.2.2}$$

If we note that

$$\frac{\partial \hat{u}_n}{\partial x} = a_{j-1}\frac{d\hat{\varphi}_{j-1}}{dx} + a_j\frac{d\hat{\varphi}_j}{dx} + a_{j+1}\frac{d\hat{\varphi}_{j+1}}{dx} \qquad x_{j-1} \leq x < x_{j+1} \tag{11.2.3}$$

then, from (1.3.3) and (1.3.4), the integral of the convection term becomes

$$-b_j\left\{ a_{j-1}\int_0^1 \frac{d\hat{\varphi}_{j-1}}{dx}\hat{\varphi}_j\,dx + a_j\int_0^1 \frac{d\hat{\varphi}_j}{dx}\hat{\varphi}_j\,dx + a_{j+1}\int_0^1 \frac{d\hat{\varphi}_{j+1}}{dx}\hat{\varphi}_j\,dx \right\}$$

$$= -b_j\left(-\frac{1}{2}a_{j-1} + \frac{1}{2}a_{j+1}\right) \tag{11.2.4}$$

Integration by parts using (11.1.2) followed by lumping of the mass eventually leads to

$$h \frac{da_j}{dt} = -\frac{\sigma}{h} (-a_{j-1} + 2a_j - a_{j+1}) - b_j\left(-\frac{1}{2} a_{j-1} + \frac{1}{2} a_{j+1}\right) + f_j$$

$$j = 1, 2, \ldots, n-1 \tag{11.2.5}$$

where a_0 and a_n are equal to 0 and

$$f_j = \int_0^1 f \hat{\varphi}_j \, dx \tag{11.2.6}$$

In the matrix form this equation can be expressed in terms of a vector a whose jth entry is a_j:

$$M \frac{da}{dt} + Ka + Ba = f \tag{11.2.7}$$

where M and K are the lumped mass matrix (8.3.3) and the stiffness matrix (8.1.14), respectively, and f is a vector whose jth entry is f_j. Here B is a matrix corresponding to the convection term defined by

$$B = \frac{1}{2} \begin{bmatrix} 0 & b_1 & & & & & \\ -b_2 & 0 & b_2 & & & & \\ & & \ddots & & & 0 & \\ & & -b_j & 0 & b_j & & \\ 0 & & & \ddots & & & \\ & & & & & & b_{n-2} \\ & & & & -b_{n-1} & 0 \end{bmatrix} \tag{11.2.8}$$

Note that this matrix is *asymmetric* and that theorems that assume the symmetry of the matrices do not apply to problems with a convection term such as (11.1.1) through (11.1.3).

If we replace the time derivative by the backward difference scheme with a time step of size Δt and set

$$a_j^k = a_j(k \, \Delta t) \tag{11.2.9}$$

$$f_j^k = f_j(k \, \Delta t) \tag{11.2.10}$$

we have from (11.2.5)

$$h \frac{a_j^{k+1} - a_j^k}{\Delta t} = -\frac{\sigma}{h} (-a_{j-1}^{k+1} + 2a_j^{k+1} - a_{j+1}^{k+1})$$

$$- b_j\left(-\frac{1}{2} a_{j-1}^{k+1} + \frac{1}{2} a_{j+1}^{k+1}\right) + f_j^{k+1} \tag{11.2.11}$$

which reduces to

$$-\left(\lambda + \frac{\Delta t}{2h}\, b_j\right) a_{j-1}^{k+1} + (1 + 2\lambda)\, a_j^{k+1} - \left(\lambda - \frac{\Delta t}{2h}\, b_j\right) a_{j+1}^{k+1} = a_j^k + \frac{\Delta t}{h} f_j^{k+1}$$

$$(11.2.12)$$

where $\lambda = \sigma\, \Delta t/h^2$.

As we have seen in the preceding sections, (11.2.12) satisfies the maximum principle if the coefficients of the first and third terms on the left-hand side of (11.2.12) are negative or zero. Thus the stability condition of scheme (11.2.12) is given by the inequality

$$\frac{2\sigma}{b_{max}} \geq h \qquad b_{max} = \max_{0 \leq x \leq 1} |b(x)| \qquad (11.2.13)$$

This means that, when the equation has a convection term, (11.2.12) is not unconditionally stable, although it is a backward scheme, and that the mesh size h of the space variable must be so small that it satisfies (11.2.13).

11.3   The Upwind Finite Element Scheme

It is natural to claim that the upper stream affects the lower stream. In fact, by taking into account more information in the upper stream than in the lower one we can improve the stability of the scheme.

In the scheme given in the previous section we approximated $\partial u/\partial x$ by (11.2.3), that is,

$$\frac{\partial \hat{u}_n}{\partial x} = \begin{cases} a_{j-1} \dfrac{d\hat{\varphi}_{j-1}}{dx} + a_j \dfrac{d\hat{\varphi}_j}{dx} & x_{j-1} \leq x < x_j \qquad (11.3.1) \\[2ex] a_j \dfrac{d\hat{\varphi}_j}{dx} + a_{j+1} \dfrac{d\varphi_{j+1}}{dx} & x_j \leq x < x_{j+1} \qquad (11.3.2) \end{cases}$$

On the other hand, we can take into account more information in the upper stream by approximating $\partial u/\partial x$ in such a way that the gradient (11.3.1) in the left interval $x_{j-1} \leq x < x_j$ is also extended to the interval $x_j \leq x < x_{j+1}$ when the flow is to the right, that is, $b_j > 0$ (Fig. 11.2). Therefore we replace $\partial u/\partial x$ by

$$a_{j-1} \frac{d\hat{\varphi}_{j-1}}{dx} + a_j \frac{d\hat{\varphi}_j}{dx} = \frac{1}{h}\,(-a_{j-1} + a_j) \qquad \text{if } b_j \geq 0$$

$$(11.3.3)$$

$$a \frac{d\hat{\varphi}_j}{dx} + a_{j+1} \frac{d\hat{\varphi}_{j+1}}{dx} = \frac{1}{h}\,(-a_j + a_{j+1}) \qquad \text{if } b_j < 0$$

in the whole interval $x_{j-1} \leq x < x_{j+1}$. Then the integral corresponding to (11.2.4) becomes

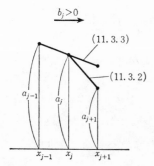

Fig. 11.2 Approximation to $\partial u / \partial x$ in the upwind finite element scheme.

$$-b_j \int_0^1 \frac{\partial \hat{u}_n}{\partial x} \hat{\varphi}_j \, dx = \begin{cases} -b_j(-a_{j-1} + a_j) & b_j \geq 0 \\ -b_j(-a_j + a_{j+1}) & b_j < 0 \end{cases} \qquad (11.3.4)$$

and the lumped mass scheme can be written as

$$-\left(\lambda + \frac{\Delta t}{h} b_j\right) a_{j-1}^{k+1} + \left(1 + 2\lambda + \frac{\Delta t}{h} b_j\right) a_j^{k+1} - \lambda a_{j+1}^{k+1}$$

$$= a_j^k + \frac{\Delta t}{h} f_j^{k+1} \qquad b_j \geq 0$$

$$\qquad (11.3.5)$$

$$-\lambda a_{j-1}^{k+1} + \left(1 + 2\lambda - \frac{\Delta t}{h} b_j\right) a_j^{k+1} - \left(\lambda - \frac{\Delta t}{h} b_j\right) a_{j+1}^{k+1}$$

$$= a_j^k + \frac{\Delta t}{h} f_j^{k+1} \qquad b_j < 0$$

This scheme is called the *upwind finite element scheme*.

The coefficients of the first and third terms on the left-hand side of (11.3.5) are always negative or zero, while the coefficients of the other terms are always positive or zero. Therefore the upwind finite element scheme is unconditionally stable.

The crucial point is that, in order for the scheme to satisfy the maximum principle, the contribution from the convection term must be included as positive or zero in the diagonal, that is, in the coefficient of a_j, while it must be included as negative or zero in the off-diagonal, that is, in the coefficient of $a_{j\pm1}$.

11.4 *A Two-Dimensional Diffusion Problem With a Convection Term*

Consider a two-dimensional diffusion problem with a convection term in a polygonal domain:

$$\frac{\partial u}{\partial t} = \sigma \left(\frac{\partial^2 u}{\partial x^2} + \frac{\partial^2 u}{\partial y^2} \right) - \left\{ b_1(x,y,t) \frac{\partial u}{\partial x} + b_2(x,y,t) \frac{\partial u}{\partial y} \right\} + f(x,y,t)$$

$$(11.4.1)$$

$$u = 0 \qquad \text{on } \partial G \tag{11.4.2}$$

$$u(x,y,0) = u_0(x,y) \tag{11.4.3}$$

The vector

$$\boldsymbol{b}(x,y,t) = (b_1(x,y,t), b_2(x,y,t)) \tag{11.4.4}$$

gives the velocity of the flow. If we use the notation

$$\text{grad } u = \nabla u = \left(\frac{\partial u}{\partial x}, \frac{\partial u}{\partial y} \right) \tag{11.4.5}$$

the second term on the right-hand side of (11.4.1) can be written as

$$-\boldsymbol{b} \cdot \text{grad } u = -\boldsymbol{b} \cdot \nabla u \tag{11.4.6}$$

We again divide the domain G into triangular subdomains of the acute type and construct a piecewise linear basis function $\hat{\phi}_i(x,y)$ corresponding to node P_i in the same way as in Sec. 9.1. Also, we employ the piecewise constant basis function $\bar{\phi}_i(x,y)$ for lumping of the mass in the barycentric region corresponding to P_i.

We write an approximate solution of this problem:

$$\hat{u}_n(x,y,t) = \sum_{j=1}^{n} a_j(t)\hat{\phi}_j(x,y) \tag{11.4.7}$$

We then substitute this expression into (11.4.1), multiply by $\hat{\phi}_i(x,y)$, and integrate. Then we have

$$\sum_{j=1}^{n} \frac{da_j}{dt} \iint_G \bar{\phi}_j \bar{\phi}_i \, dx \, dy + \sigma \sum_{j=1}^{n} a_j \iint_G \left(\frac{\partial \hat{\phi}_j}{\partial x} \frac{\partial \hat{\phi}_i}{\partial x} + \frac{\partial \hat{\phi}_j}{\partial y} \frac{\partial \hat{\phi}_i}{\partial y} \right) dx \, dy$$

$$+ \iint_G b_1 \frac{\partial \hat{u}_n}{\partial x} \hat{\phi}_i \, dx \, dy + \iint_G b_2 \frac{\partial \hat{u}_n}{\partial y} \hat{\phi}_i \, dx \, dy = \iint_G f \hat{\phi}_i \, dx \, dy$$

$$(11.4.8)$$

The lumping of the mass (9.2.2) is carried out in the first term on the left-hand side. The first and second terms on the left-hand side can be treated as before. Special attention is required when we handle the third and fourth terms that correspond to the convection term.

11.5 Upwind Finite Elements

Based on the analysis in Sec. 11.3 for a one-dimensional problem, we see that the scheme obtained by discretizing time in (11.4.8) satisfies the maximum principle if the coefficient corresponding to the convection term is positive or zero in the diagonal term of the scheme and negative or zero in the off-diagonal term. In order to realize this situation we introduce the *upwind finite element* corresponding to node P_i. Let the coordinates of P_i be (x_i, y_i). The x upwind finite element τ_x^i for P_i is a triangle that the line parallel to the x axis passing through P_i traverses on the left side of P_i when $b_1(x_i, y_i, t) \geq 0$ and on the right side of P_i when $b_1(x_i, y_i, t) < 0$ (Fig. 11.3). When the line coincides with the common side of two triangles, either triangle may be chosen for τ_x^i. We define the y upwind finite element τ_y^i in the same way.

Let the nodes other than P_i of the x upwind finite element τ_x^i be P_j and P_k. Here P_i, P_j, and P_k should be arranged in a positive direction, that is, counterclockwise. In τ_x^i, $\hat{u}_n(x, y, t)$ can be expressed as

$$\hat{u}_n(x, y, t) \Big|_{\tau_x^i} = a_i(t)\xi_i(x, y) + a_j(t)\xi_j(x, y) + a_k(t)\xi_k(x, y)$$

(11.5.1)

where ξ_i, ξ_j, and ξ_k are the shape functions corresponding to P_i, P_j, and P_k, respectively. Therefore, from (5.3.5) we have

$$\frac{\partial \hat{u}_n}{\partial x} \Big|_{\tau_x^i} = \frac{1}{2S_x^i} \left[(y_j - y_k)a_i + (y_k - y_i)a_j + (y_i - y_j)a_k \right]$$

(11.5.2)

where S_x^i is the area of τ_x^i and is positive. Suppose that $b_1(x_i, y_i, t) \geq 0$. Then, from the definition of τ_x^i we see that

$$y_j - y_k > 0 \qquad y_k - y_i \leq 0 \qquad y_i - y_j \leq 0 \qquad (11.5.3)$$

These inequalities may be understood intuitively if we think of the meaning of $\partial \xi_i / \partial x$, $\partial \xi_j / \partial x$, and $\partial \xi_k / \partial x$. Multiplying by b_1 we have

$$b_1(y_j - y_k) \geq 0 \qquad b_1(y_k - y_i) \leq 0 \qquad b_1(y_i - y_j) \leq 0$$

(11.5.4)

When $b_1(x_i, y_i, t) < 0$, inequalities opposite those in (11.5.3) hold. Thus

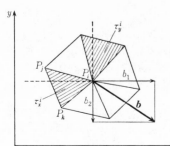

Fig. 11.3   Upwind finite elements.

the coefficient of the diagonal term a_i is positive or zero, and those of the off-diagonal terms a_j and a_k are negative or zero.

11.6   *The Two-Dimensional Finite Element Scheme*

Now we compute the third term on the left-hand side of (11.4.8) explicitly. The idea of the upwind approximation is to use the information in the upwind finite element τ_x^i over the whole range consisting of all the triangular elements around node P_i, for example, over the hexagon shown in Fig. 11.3. First we replace $b_1(x,y,t)$ by $b_1(x_i,y_i,t)$ at node P_i. Then $b_1(\partial \hat{u}_n/\partial x)$ can be moved outside the integral sign because it does not depend on x and y, and we have only to integrate $\hat{\phi}_i$. From (5.6.3) the integral of $\hat{\phi}_i$ in each triangular element around node P_i is one-third the area of each triangular element. Summing up the contributions from the triangular elements around P_i, we eventually have for the third term on the left-hand side of (11.4.8)

$$\iint_G b_1 \frac{\partial \hat{u}_n}{\partial x} \hat{\phi}_i \, dx \, dy \doteq \frac{1}{6} b_1(x_i,y_i,t) \frac{S^i}{S_x^i} [(y_j - y_k)a_i$$
$$+ (y_k - y_i)a_j + (y_i - y_j)a_k] \qquad (11.6.1)$$

where S^i is the area of the barycentric region corresponding to P_i. Similarly, we approximate the fourth term on the left-hand side of (11.4.8) as

$$\iint_G b_2 \frac{\partial \hat{u}_n}{\partial y} \hat{\phi}_i \, dx \, dy \doteq \frac{1}{6} b_2 (x_i,y_i,t) \frac{S^i}{S_y^i} [(x_j - x_k)a_i$$
$$+ (x_k - x_i)a_j + (x_i - x_j)a_k] \qquad (11.6.2)$$

Next we discretize time, that is, we replace the time derivative by the time difference. We let the time mesh be Δt as before and employ the backward difference. The result is

$$M \frac{a((k+1)\,\Delta t) - a(k\,\Delta t)}{\Delta t} + Ka((k+1)\,\Delta t) + Ba((k+1)\,\Delta t)$$

$$= f((k+1)\,\Delta t) \qquad (11.6.3)$$

where M and K are given by (9.2.4) and (9.2.5), respectively, and f is a vector whose jth entry is $f_j((k+1)\,\Delta t)$, the value of the right-hand side of (11.4.8) at $t = (k+1)\,\Delta t$. Here B is the matrix corresponding to the convection term whose entries are easily obtained from (11.6.1) and (11.6.2). Note that matrix B is again asymmetric.

Starting from the initial value $a(0)$ we iterate this scheme and obtain a finite element solution $\hat{u}_n(x,y,t)$.

11.7 *Stability of the Scheme*

We multiply both sides of scheme (11.6.3) by Δt and write down its ith row explicitly. Then, if we move $-M_{ii}a_i(k\,\Delta t)$ and the off-diagonal terms to the right-hand side, we obtain

$$(M_{ii} + \Delta t\, K_{ii} + \Delta t\, B_{ii})\, a_i((k+1)\,\Delta t)$$

$$= M_{ii}a_i(k\,\Delta t) - \Delta t \sum_{\substack{j=1 \\ j\neq i}}^{n} K_{ij}a_j((k+1)\,\Delta t)$$

$$- \Delta t \sum_{\substack{j=1 \\ J\neq i}}^{n} B_{ij}a_j((k+1)\,\Delta t) + f_i((k+1)\,\Delta t)$$

$$(11.7.1)$$

It is evident that $m_{ii} > 0$ and $K_{ii} > 0$. Also, if the triangulation is of the acute type we have $K_{ij} \leq 0$ from (9.3.6). On the other hand, from (11.5.4) we see that

$$B_{ii} \geq 0 \qquad B_{ij} \leq 0 \qquad j \neq i \qquad\qquad (11.7.2)$$

as long as we employ the upwind finite element. Therefore, if the triangulation is of the acute type, the coefficients of a_i and a_j are all positive or zero, and we conclude that scheme (11.6.3) satisfies the maximum principle; that is, (11.6.3) is unconditionally stable.

12

A Free Boundary Problem

12.1 The Stefan Problem

In the problems considered so far the domain G does not change with time. On the other hand, there are many natural phenomena in which the shape of the domain changes with time according to the physical state of its interior. The change in the boundary between water and ice resulting from melting of the ice is a typical example. A problem obtained by formulating such a phenomenon mathematically is called a *free boundary problem*.

As an example of a free boundary problem we consider here the *Stefan problem*. For simplicity we take the problem in one space dimension:

$$\frac{\partial u}{\partial t} = \sigma \frac{\partial^2 u}{\partial x^2} \qquad 0 < x < s(t) \qquad 0 < t \le T \qquad (12.1.1)$$

$$u(0, t) = g(t) \qquad (12.1.2)$$

$$u(s(t), t) = 0 \qquad (12.1.3)$$

$$u(x, 0) = f(x) \qquad 0 \le x \le b = s(0) \qquad 0 < b \qquad (12.1.4)$$

where $g(t)$ and $f(x)$ are the given initial functions and $b = s(0)$ is also assumed to be given. For example, let u be the temperature of the water and $x = s(t)$ describe the boundary between the water and the ice. The water exists at $0 < x < s(t)$ and, warming up the boundary at $x = 0$ to temperature $g(t)$ causes the ice to melt at the other boundary $x = s(t)$ and bound-

138

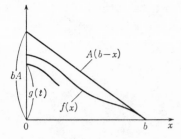

Fig. 12.1 Assumptions for the initial and the boundary conditions.

ary $s(t)$ moves toward the right. The boundary then changes according to the Stefan condition

$$\frac{ds}{dt} = -\kappa \frac{\partial u}{\partial x}(s(t), t) \qquad 0 < t \leq T \qquad (12.1.5)$$

where $\kappa > 0$ is the thermal conductivity and the right-hand side represents the heat flowing into the boundary from the water. Therefore (12.1.5) describes the latent heat; that is, if the heat that amounts to the right-hand side is given per unit time, then ds of ice melts during time dt. In short, the speed of the boundary motion when the ice melts is proportional to the temperature gradient of the water at the boundary. The objective of this problem is to make the temperature $u(x, y)$ coincide with the position of the boundary $s(t)$.

For later convenience, we make the assumption

$$0 \leq f(x) \leq A(b - x) \qquad (12.1.6)$$

$$0 < g(t) \leq bA \qquad (12.1.7)$$

(Fig. 12.1), where A is a constant describing the condition that the initial and the boundary values do not increase too much.

12.2 *Time-Dependent Basis Functions*

We apply the FEM to the Stefan problem in one space dimension. Since the boundary moves as time elapses, we use subdomains that change with time. We compute an approximate location of the boundary together with an approximate solution of u, so that we write it as $s_n(t)$ instead of $s(t)$ in order to show explicitly that it is an approximation. First we fix time t and divide the domain $0 \leq x \leq s_n(t)$ at t into n equal subintervals. Next, at each node

$$x_j = jh \qquad h = h(t) = \frac{s_n(t)}{n} \qquad (12.2.1)$$

we construct a piecewise linear basis function $\hat{\phi}_j(x,t)$ (Fig. 12.2):

$$\hat{\varphi}_j(x,t) = \begin{cases} 0 & 0 \leq x < x_{j-1} \\[2mm] \dfrac{x - x_{j-1}}{x_j - x_{j-1}} & x_{j-1} \leq x < x_j \\[2mm] \dfrac{x_{j+1} - x}{x_{j+1} - x_j} & x_j \leq x < x_{j+1} \\[2mm] 0 & x_{j+1} \leq x \leq s_n(t) \end{cases} \tag{12.2.2}$$

For $\hat{\phi}_0$ and $\hat{\phi}_n$ we take the right half and the left half of (12.2.2), respectively. Although the shape of the basis function $\hat{\phi}_j(x,t)$ is similar to that of functions used in the problems considered so far, we should note that it depends on time t through $x_j = jh(t)$. The time derivative of $\hat{\phi}_j(x,t)$ becomes

$$\frac{\partial \hat{\varphi}_j}{\partial t} = \begin{cases} 0 & 0 \leq x < x_{j-1} \\[2mm] -\dfrac{(x - x_{j-1})\dot{x}_j + (x_j - x)\dot{x}_{j-1}}{(x_j - x_{j-1})^2} & x_{j-1} \leq x < x_j \\[2mm] \dfrac{(x - x_j)\dot{x}_{j+1} + (x_{j+1} - x)\dot{x}_j}{(x_{j+1} - x_j)^2} & x_j \leq x < x_{j+1} \\[2mm] 0 & x_{j+1} \leq x \leq s_n(t) \end{cases} \tag{12.2.3}$$

where $\dot{x}_j$ represents the derivative with respect to time t:

$$\dot{x}_j = \frac{dx_j}{dt} = j\frac{dh}{dt} = \frac{j}{n}\frac{ds_n}{dt} \tag{12.2.4}$$

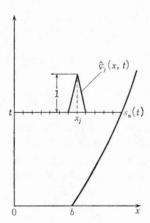

Fig. 12.2 Time-dependent basis function $\hat{\phi}_j(x,t)$.

12.3 *Application of the Finite Element Method*

An approximate solution of the problem (12.1.1) through (12.1.5) can be written in terms of the basis functions defined in the preceding section:

$$\hat{u}_n(x,t) = \sum_{j=0}^{n} a_j(t)\hat{\varphi}_j(x,t) \tag{12.3.1}$$

From the boundary condition (12.1.2) and (12.1.3) we have

$$a_0(t) = g(t)$$
$$a_n(t) = 0 \tag{12.3.2}$$

Although $a_n(t) = 0$, we add the term for $j = n$ for later convenience. We substitute $\hat{u}_n$ in (12.3.1) for u in (12.1.1). Multiplying both sides by $\hat{\phi}_k(x,t)$, $k = 1, 2, \ldots, n-1$, and integrating by parts yield

$$M(t)\frac{d\boldsymbol{a}}{dt} + \{K(t) + N(t)\}\boldsymbol{a} = 0 \tag{12.3.3}$$

where

$$\boldsymbol{a}(t) = \begin{pmatrix} a_0(t) \\ a_1(t) \\ \vdots \\ a_n(t) \end{pmatrix} \tag{12.3.4}$$

and where M and K are $(n-1) \times (n-1)$ matrices whose ij entries are given by

$$M_{ij} = \begin{cases} h(t) & j = i \\ 0 & j \neq i \end{cases} \tag{12.3.5}$$

$$K_{ij} = \begin{cases} -\dfrac{\sigma}{h(t)} & j = i-1 \\[2mm] \dfrac{2\sigma}{h(t)} & j = i \\[2mm] -\dfrac{\sigma}{h(t)} & j = i+1 \\[2mm] 0 & j \neq i-1, i, i+1 \end{cases} \tag{12.3.6}$$

respectively. Lumping is performed in the mass matrix M. The matrix N is an $(n-1) \times (n-1)$ matrix that comes from the derivative of $\hat{\phi}_j(x,t)$ with respect to time:

$$N_{ij} = \begin{cases} \dfrac{1}{6}(3i-1)\dfrac{dh}{dt} & j=i-1 \\[2ex] \dfrac{1}{3}\dfrac{dh}{dt} & j=i \\[2ex] -\dfrac{1}{6}(3i+1)\dfrac{dh}{dt} & j=i+1 \\[2ex] 0 & j \neq i-1, i, i+1 \end{cases} \qquad (12.3.7)$$

Note that it is not symmetric.

Next we replace the time derivative in the ordinary differential equation (12.3.3) by the backward time difference. Time t is discretized with a uniform mesh Δt, and for M, K, and N we use the values at time $t = (k+1)\,\Delta t$:

$$M\frac{a((k+1)\,\Delta t) - a(k\,\Delta t)}{\Delta t} + (K+N)a((k+1)\,\Delta t) = 0$$
$$(12.3.8)$$

Since time t has an upper limit, we have

$$m\,\Delta t = T \qquad (12.3.9)$$

The increment Δs_n in the position $s_n(t)$ as time elapses from $t = k\,\Delta t$ to $t = (k+1)\,\Delta t$ is computed by replacing (12.1.5) by

$$\frac{\Delta s_n}{\Delta t} = -\kappa\frac{\partial \hat{u}_n}{\partial x}(s_n((k+1)\,\Delta t),\ (k+1)\,\Delta t) = \kappa\frac{na_{n-1}((k+1)\,\Delta t)}{s_n((k+1)\,\Delta t)}$$
$$(12.3.10)$$

We summarize here the procedure shown above, denoting the approximate solution by $a_j^k = a_j(k\,\Delta t)$. First we compute the initial value:

$$a_j^0 = f(x_j) \qquad j = 0, 1, \ldots, n$$

$$\Delta s_n\,\Delta t = \kappa\frac{na_{n-1}^0}{b}\,\Delta t \qquad (12.3.11)$$

$$s_n\,\Delta t = b + \Delta s_n\,\Delta t$$

Then we repeat the following for $k = 0, 1, \ldots, m-1$. In the first stage the system of linear equations

$$\{M((k+1)\,\Delta t) + \Delta t\ (K((k+1)\,\Delta t)$$
$$+ N((k+1)\,\Delta t))\}a((k+1)\,\Delta t) = Ma(k\,\Delta t)$$
$$(12.3.12)$$

which is obtained by multiplying (12.3.8) by Δt, is solved for $a((k+1)\,\Delta t)$. In this system of equations we compute M, K, and N

using $s_n((k+1)\,\Delta t) = nh((k+1)\,\Delta t)$ and $\Delta s_n((k+1)\,\Delta t)$. In the second stage we compute

$$\Delta s_n((k+2)\,\Delta t) = \kappa \, \frac{na_{n-1}^{k+1}}{s_n((k+1)\,\Delta t)} \, \Delta t \qquad (12.3.13)$$

$$s_n((k+2)\,\Delta t) = s_n((k+1)\,\Delta t) + \Delta s_n((k+2)\,\Delta t)$$
$$(12.3.14)$$

using the $(n-1)$th entry a_{n-1}^{k+1} of the solution $a((k+1)\,\Delta t)$.

By the procedure stated above we obtain an approximate solution $\hat{u}_n(x,t)$ and an approximate location of the boundary. Note that this scheme is nonlinear with respect to a_j^k.

12.4 *Stability of the Scheme*

In this section we discuss the stability of the scheme given in the previous section. Let

$$\alpha_k = \frac{\sigma n^2 \,\Delta t}{s_n^2(k\,\Delta t)} + \frac{\Delta s_n(k\,\Delta t)}{6 s_n(k\,\Delta t)} \qquad (12.4.1)$$

$$\beta_k = \frac{\Delta s_n(k\,\Delta t)}{2 s_n(k\,\Delta t)} \qquad (12.4.2)$$

in order to simplify the representation of the scheme. Then the jth row of (12.3.12) becomes

$$-(\alpha_{k+1} - j\beta_{k+1})\,a_{j-1}^{k+1} + (1 + 2\alpha_{k+1})\,a_j^{k+1} - (\alpha_{k+1} + j\beta_{k+1})\,a_{j+1}^{k+1} = a_j^k$$
$$j = 1, 2, \ldots, n-1 \qquad (12.4.3)$$

where $a_0^{k+1} = g((k+1)\,\Delta t)$ and $a_n^{k+1} = 0$ are known. As we have seen in the preceding sections, if the coefficient of the second term on the left-hand side is positive and those of the first and second terms are negative or zero, scheme (12.4.3) is stable; that is, it satisfies the maximum principle. Therefore, if both

$$\Delta s_n(k\,\Delta t) \geq 0 \qquad (12.4.4)$$

$$n\beta_k \leq \alpha_k \qquad (12.4.5)$$

are always satisfied, (12.4.3) is stable.

In order for (12.4.4) to hold, that is, in order for the boundary always to move to the right as time elapses, it suffices to show that $a_{n-1}^k > 0$. On the other hand, (12.4.5) holds if Δs_n is not too large compared with s_n. In other words, it suffices for the gradient of $\hat{u}_n$ at the boundary to be bounded moderately. Thus for A in (12.1.6) and (12.1.7) we assume

$$A \leq \frac{2\sigma n}{\kappa l} \qquad l = b + \kappa A T \tag{12.4.6}$$

and prove

$$0 \leq \frac{a_j^k}{(1 - j/n)\, s_n(k\, \Delta t)} \leq A \qquad j = 0, 1, \ldots, n-1$$
$$k = 0, 1, \ldots, m \tag{12.4.7}$$

If this inequality holds, the gradient of $\hat{u}_n$ at the boundary is bounded by a constant A, as is evident from Fig. 12.3; hence the boundary does not move too quickly to the right and the maximum principle holds.

To prove the inequality (12.4.7) we let

$$d_j^k = A\left(1 - \frac{j}{n}\right) s_n(k\, \Delta t) - a_j^k \tag{12.4.8}$$

and show that

$$0 < b \leq s_n\, \Delta t \leq s_n(2\, \Delta t) \leq \cdots \leq s_n(k\, \Delta t) \tag{12.4.9}$$

$$0 \leq a_j^k \tag{12.4.10}$$

$$0 \leq d_j^k \tag{12.4.11}$$

To this end we solve (12.4.8) for a_j^k and substitute it into (12.4.3), which leads to the following scheme satisfied by d_j^k:

$$-(\alpha_{k+1} - j\beta_{k+1})\, d_{j-1}^{k+1} + (1 + 2\alpha_{k+1})\, d_j^{k+1} - (\alpha_{k+1} + j\beta_{k+1})\, d_{j+1}^{k+1}$$
$$= d_j^k + \frac{j}{n}\, A\, \Delta s_n\, ((k+1)\, \Delta t) \tag{12.4.12}$$

Now we prove (12.4.9) through (12.4.11) by induction with respect to

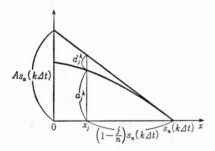

Fig. 12.3 The condition $0 \leq \dfrac{a_j^k}{(1 - j/n)\, s_n(k\, \Delta t)} \leq A$.

k. First it is evident that (12.4.9) through (12.4.11) hold for $k = 0$ because of (12.1.4), (12.1.6), and (12.3.11). Next assume that (12.4.9) through (12.4.11) hold for k. Then it is evident that, from (12.3.13), $\Delta s_n ((k + 1) \Delta t) \geq 0$; that is, $s_n(k \Delta t) \leq s_n ((k + 1) \Delta t)$ holds. Also, from $0 \leq d_{n-1}^k$ and (12.3.13)

$$\frac{\Delta s_n ((k + 1) \Delta t)}{\Delta t} = \kappa \frac{n a_{n-1}^k}{s_n(k \Delta t)} \leq \kappa A \qquad (12.4.13)$$

holds. On the other hand, from

$$s_n((k + 1) \Delta t) \leq b + (k + 1) \Delta t \kappa A \leq b + \kappa A T = l \qquad (12.4.14)$$

we have

$$\alpha_{k+1} - n\beta_{k+1} = \frac{\Delta t}{s_n((k + 1) \Delta t)} \left\{ \frac{\sigma n^2}{s_n((k + 1) \Delta t)} \right.$$
$$\left. - \frac{3n - 1}{6} \frac{\Delta s_n ((k + 1) \Delta t)}{\Delta t} \right\} \geq \frac{\kappa n \Delta t}{2l} \left(\frac{2\sigma n}{\kappa l} - A \right) \geq 0 \qquad (12.4.15)$$

Therefore the maximum principle holds for (12.4.3); that is, the first inequality in (8.6.5) holds for $\theta = 1$ and $f_j^k = 0$, and, if we note that $a_0^{k+1} = g((k + 1) \Delta t) \geq 0$ and $a_{n+1}^{k+1} = 0$, we eventually obtain

$$a_j^{k+1} \geq 0 \qquad (12.4.16)$$

In a similar way we can prove that the maximum principle holds for (12.4.12); hence if we replace $\Delta t f_j^k$ by $\Delta s_n ((k + 1) \Delta t)$ in (8.6.5) and note that $d_0^{k+1} = As_n((k+1) \Delta t) - a_0^{k+1} \geq Ab - g((k+1) \Delta t) \geq 0$ from (12.1.7) and that $d_n^{k+1} = 0$, we conclude that

$$d_j^{k+1} \geq 0 \qquad (12.4.17)$$

Thus (12.4.9) through (12.4.11) are proved.

As seen above, the Stefan problem is a nonlinear one because the solution $\hat{u}_n(x,t)$ and the location of the boundary $s_n(t)$ interact with each other. Thus a procedure more complicated than the one presented above is necessary to prove the stability of the scheme than in the case of a linear problem.

13

Nonlinear Problems and Iterative Approximation

13.1 *A Nonlinear Problem and an Equation of the Weak Form*

Although almost all the problems discussed so far have been linear with respect to the unknown function, the FEM can also be applied to various kinds of nonlinear problems. In this chapter we return to time-independent problems and apply the FEM to typical nonlinear problems.

If we have a weak form of the given differential equation on hand, we can usually apply the FEM. The following equation defined in a two-dimensional domain G is a typical example to which the FEM can be applied:

$$-\left\{\frac{\partial}{\partial x}\left(\alpha(u)\,\frac{\partial u}{\partial x}\right)+\frac{\partial}{\partial y}\left(\beta(u)\,\frac{\partial u}{\partial y}\right)\right\}=f(x,y) \qquad (13.1.1)$$

$$u=0 \qquad \text{on } \partial G \qquad (13.1.2)$$

where α and β are the functions of u, hence (13.1.1) is a nonlinear equation with respect to u. If we multiply both sides of (13.1.1) by v belonging to $\mathring{H}_1$ as defined in Chap. 4 and integrate, we have the following equation in the weak form:

$$\iint_G \left\{\alpha(u)\,\frac{\partial u}{\partial x}\,\frac{\partial v}{\partial x}+\beta(u)\,\frac{\partial u}{\partial y}\,\frac{\partial v}{\partial y}\right\}dx\,dy=\iint_G fv\,dx\,dy$$
$$(13.1.3)$$

$$u = 0 \qquad \text{on } \partial G \qquad (13.1.4)$$

We express an approximate solution in terms of a linear combination of the piecewise linear functions $\{\hat{\phi}_j\}$ defined in Sec. 4.5:

$$\hat{u}_n(x, y) = \sum_{j=1}^{n} a_j \hat{\varphi}_j(x, y) \qquad (13.1.5)$$

and chose $\hat{\phi}_k$, $k = 1, 2, \ldots, n$, for v. Then we obtain from (13.1.3) an approximate equation

$$\iint_G \left\{ \alpha(\hat{u}_n) \frac{\partial \hat{u}_n}{\partial x} \frac{\partial \hat{\varphi}_k}{\partial x} + \beta(\hat{u}_n) \frac{\partial \hat{u}_n}{\partial y} \frac{\partial \hat{\varphi}_k}{\partial y} \right\} dx \, dy = \iint_G f \hat{\varphi}_k \, dx \, dy$$

$$(13.1.6)$$

The coefficients include $\alpha(\hat{u}_n)$ and $\beta(\hat{u}_n)$, so that this equation is nonlinear with respect to the unknowns $\{a_j\}$. However, if the dependency of u on α and β is small, we can treat these terms as approximately linear. In order to solve this kind of nonlinear equation we usually apply an iterative method as shown below.

An *iterative method* is a method in which a sequence of approximate functions $\hat{u}_n$ are improved by iterative substitution. Thus a superscript (m) is attached to $\hat{u}_n$ corresponding to the mth approximation

$$\hat{u}_n^{(m)}(x, y) = \sum_{j=1}^{n} a_j^{(m)} \hat{\varphi}_j(x, y) \qquad (13.1.7)$$

and we construct the following iterative scheme from (13.1.6):

$$\iint_G \left\{ \alpha(\hat{u}_n^{(m)}) \frac{\partial \hat{u}_n^{(m+1)}}{\partial x} \frac{\partial \hat{\varphi}_k}{\partial x} + \beta(u_n^{(m)}) \frac{\partial \hat{u}_n^{(m+1)}}{\partial y} \frac{\partial \hat{\varphi}_k}{\partial y} \right\} dx \, dy = \iint_G f \hat{\varphi}_k \, dx \, dy$$

$$m = 0, 1, 2, \ldots \qquad (13.1.8)$$

If we write this equation explicitly in terms of the unknowns $\{a_j^{(m+1)}\}$, it becomes

$$\sum_{j=1}^{n} a_j^{(m+1)} \iint_G \left\{ \alpha(\hat{u}_n^{(m)}) \frac{\partial \hat{\varphi}_j}{\partial x} \frac{\partial \hat{\varphi}_k}{\partial x} + \beta(\hat{u}_n^{(m)}) \frac{\partial \hat{\varphi}_j}{\partial y} \frac{\partial \hat{\varphi}_k}{\partial y} \right\} dx \, dy$$

$$= \iint_G f \hat{\varphi}_k \, dx \, dy \qquad m = 0, 1, 2, \ldots$$

$$(13.1.9)$$

Then the procedure of iteration is as follows. First we choose an appropriate initial value $\{a_j^{(0)}\}$, construct $\hat{u}_n^{(0)}$, and compute $\alpha(\hat{u}_n^{(0)})$ and $\beta(\hat{u}_n^{(0)})$. Then (13.1.9) becomes an equation with respect to $\hat{u}_n^{(1)}$, that is, a system of linear equations with respect to $\{a_j^{(1)}\}$. We solve it for $\{a_j^{(1)}\}$ and compute $\alpha(\hat{u}_n^{(1)})$ and $\beta(\hat{u}_n^{(1)})$ using the solution. Then (13.1.9) again becomes a system of linear equations with respect to $\{a_j^{(2)}\}$, and we solve it for $\{\hat{u}_n^{(2)}\}$.

We repeat this operation until convergence appears to be attained. Explicitly, we set an error tolerance ϵ in advance and repeat the iteration until

$$|a_j^{(m+1)} - a_j^{(m)}| < \epsilon \qquad j = 1, 2, \ldots, n \qquad (13.1.10)$$

holds. It must be noted that whether $\hat{u}_n^{(m)}$ converges to an appropriate approximate solution of u, or whether the system of linear equations with respect to the unknowns $\{a_j^{(m)}\}$ has a solution as m is increased, depends on the nature of α and β.

13.2 The Navier-Stokes Equation and Its Weak Form

As a typical example of application of the iterative method to the weak form corresponding to the given equation, we consider the Navier-Stokes equation for an incompressible viscous fluid in two dimensions.

We denote the flow velocity and pressure by

$$u = u(x, y) = (u_1(x, y), u_2(x, y)) \qquad (13.2.1)$$

$$p = p(x, y) \qquad (13.2.2)$$

respectively. Then the Navier-Stokes equation in a two-dimensional domain G can be written as

$$\rho u \cdot \operatorname{grad} u_1 + \frac{\partial p}{\partial x} - \mu \, \Delta u_1 = \rho f_1 \qquad (13.2.3)$$

$$\rho u \cdot \operatorname{grad} u_2 + \frac{\partial p}{\partial y} - \mu \, \Delta u_2 = \rho f_2 \qquad (13.2.4)$$

where ρ = density of fluid
$\quad \mu$ = viscosity coefficient
$\quad f = (f_1, f_2)$ = body force, for example, the force of gravity

The velocity u satisfies the following *equation of continuity:*

$$\operatorname{div} u = 0 \qquad (13.2.5)$$

The first terms on the left-hand side of (13.2.3) and (13.2.4) are the nonlinear terms under consideration. If the contribution of these terms to the equation is supposed to be small, we can apply the iterative method presented in the previous section.

Suppose that the boundary ∂G consists of two parts ∂G_1 and ∂G_2. In the first part ∂G_1 the velocity is assigned, while in the second part ∂G_2 the force $r = (r_1, r_2)$ is assigned (Fig. 13.1). That is, the following assumption is made for the boundary condition on ∂G_1 and ∂G_2:

$$u = g = (g_1, g_2) \qquad \text{on } \partial G_1 \qquad (13.2.6)$$

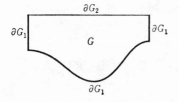

Fig. 13.1 The boundaries ∂G_1 and ∂G_2.

$$\left(-p + \mu\, \frac{\partial u_1}{\partial x}\right) \frac{\partial x}{\partial n} + \mu\, \frac{\partial u_1}{\partial y}\, \frac{\partial y}{\partial n} = r_1 \tag{13.2.7}$$

on ∂G_2

$$\mu\, \frac{\partial u_2}{\partial x}\, \frac{\partial x}{\partial n} + \left(-p + \mu\, \frac{\partial u_2}{\partial y}\right) \frac{\partial y}{\partial n} = r_2 \tag{13.2.8}$$

where $\partial/\partial n$ is the derivative of the outward normal on ∂G_2. The first boundary condition (13.2.6) corresponds to the case where the velocity at an entrance to the domain is prescribed, or to the case where, if there is a wall, the velocity component perpendicular to the wall is assigned as 0. The second boundary condition (13.2.7) and (13.2.8) corresponds to the condition at the surface of water where the atmospheric pressure is prescribed, for example. Although it may sometimes be natural to assign the second boundary condition to the two directions parallel to and perpendicular to the boundary surface, here we assign the condition to the directions of the x and y axes.

We denote by H_1^* the space of functions that are differentiable in the domain G and vanish on the boundary ∂G_1, as in Sec. 4.6. Nothing is assigned on ∂G_2. Then we multiply both sides of (13.2.3) and (13.2.4) by an arbitrary function $v \in H_1^*$ and integrate by parts in G, obtaining a weak form:

$$\iint_G \left(\rho u_1 \frac{\partial u_1}{\partial x} + \rho u_2 \frac{\partial u_1}{\partial y} + \frac{\partial p}{\partial x} - \mu\, \Delta u_1 - \rho f_1 \right) v \; dx \; dy$$

$$= \iint_G \left\{ \left(\rho u_1 \frac{\partial u_1}{\partial x} + \rho u_2 \frac{\partial u_1}{\partial y}\right) v + \left(-p + \mu\, \frac{\partial u_1}{\partial x}\right) \frac{\partial v}{\partial x} + \mu\, \frac{\partial u_1}{\partial y}\, \frac{\partial v}{\partial y} \right\} dx \; dy$$

$$- \int_{\partial G_1 + \partial G_2} \left\{ \left(-p + \mu\, \frac{\partial u_1}{\partial x}\right) \frac{\partial x}{\partial n} + \mu\, \frac{\partial u_1}{\partial y}\, \frac{\partial y}{\partial n} \right\} v \; d\sigma$$

$$- \iint_G \rho f_1 v \; dx \; dy = 0 \tag{13.2.9}$$

When integrating by parts we used an identity

$$\iint_G \frac{\partial}{\partial x} (UV) \; dx \; dy = \iint_G \frac{\partial U}{\partial x} V \; dx \; dy + \iint_G U \frac{\partial V}{\partial x} \; dx \; dy$$

$$= \int_{\partial G} UV \frac{\partial x}{\partial n} \; d\sigma \tag{13.2.10}$$

and a similar identity with respect to y. Taking into account the boundary condition (13.2.7) and the fact that $v = 0$ on ∂G_1, we eventually obtain the following equation corresponding to the weak form of (13.2.3):

$$\iint_G \left\{ \left(\rho u_1 \frac{\partial u_1}{\partial x} + \rho u_2 \frac{\partial u_1}{\partial y} \right) v + \left(-p + \mu \frac{\partial u_1}{\partial x} \right) \frac{\partial v}{\partial x} + \mu \frac{\partial u_1}{\partial y} \frac{\partial v}{\partial y} \right\} dx \, dy$$

$$= \int_{\partial G_2} r_1 v \, d\sigma + \iint_G \rho f_1 v \, dx \, dy \qquad \forall v \in H_1^* \qquad (13.2.11)$$

Similarly, we have the following weak form corresponding to (13.2.4):

$$\iint_G \left\{ \left(\rho u_1 \frac{\partial u_2}{\partial x} + \rho u_2 \frac{\partial u_2}{\partial y} \right) v + \mu \frac{\partial u_2}{\partial x} \frac{\partial v}{\partial x} + \left(-p + \mu \frac{\partial u_2}{\partial y} \right) \frac{\partial v}{\partial y} \right\} dx \, dy$$

$$= \int_{\partial G_1} r_2 v \, d\sigma + \iint_G \rho f_2 v \, dx \, dy \qquad \forall v \in H_1^* \qquad (13.2.12)$$

In order to derive a weak form of the equation of continuity (13.2.5), we multiply both sides by an arbitrary function $q \in H_0$ and integrate:

$$\iint_G \left(\frac{\partial u_1}{\partial x} + \frac{\partial u_2}{\partial y} \right) q \, dx \, dy = 0 \qquad \forall q \in H_0 \qquad (13.2.13)$$

where H_0 is a space of functions that are integrable in G and nothing is prescribed on the boundary and q is a function corresponding to the pressure.

13.3 The Finite Element Solution of the Navier-Stokes Equation

In order to apply the FEM to weak form equations (13.2.11) through (13.2.13), we approximate the velocity (u_1, u_2) of the flow by $(\hat{u}_1, \hat{u}_2)$ in terms of the basis functions $\{\hat{\phi}_j\}$:

$$\hat{u}_1(x, y) = \sum_j a_j \hat{\varphi}_j(x, y) \qquad (13.3.1)$$

$$\hat{u}_2(x, y) = \sum_j b_j \hat{\varphi}_j(x, y) \qquad (13.3.2)$$

On the other hand, for the pressure p we do not use the basis functions used for the velocity (u_1, u_2). We represent an approximation $\hat{p}$ to p in terms of basis functions $\{\hat{\psi}_j\}$ that differ from $\{\hat{\phi}_j\}$:

$$\hat{p}(x, y) = \sum_i c_j \hat{\psi}_j(x, y) \qquad (13.3.3)$$

Although it is sufficient to assume first-order differentiability of $\{\hat{\phi}_j\}$

and only integrability of $\{\hat{\psi}_j\}$, as seen from the order of derivatives appearing in (13.2.11) through (13.2.13), from the theoretical point of view there is another kind of restriction on the pair $\{\hat{\phi}_j\}$ and $\{\hat{\psi}_j\}$. To achieve improved accuracy in some practical problems, it is appropriate, for example, to employ the piecewise polynomials of second order described in Sec. 7.2 for $\{\hat{\phi}_j\}$ and piecewise linear functions for $\{\hat{\psi}_j\}$.

We substitute the approximation (13.3.1) and (13.3.2) into (13.2.11) through (13.2.13) and choose $\hat{\phi}_k$ for v and $\hat{\psi}_k$ for q. The reason why we choose $\hat{\psi}_n$ for q will be evident if we notice the correspondence of the terms $-p \, (\partial v/\partial x)$ in (13.2.11) and $-p \, (\partial v/\partial y)$ in (13.2.12) to the integrand in (13.2.13). Then we attach the superscript (m) to u_1 and u_2 in the nonlinear terms that do not include derivatives and attach the superscript $(m + 1)$ to all the other terms, which results in an iterative scheme as stated in the previous section. That is, the scheme for computing $\{a_j^{(m+1)}, b_j^{(m+1)}, c_j^{(m+1)}\}$ at the $(m + 1)$th step using $\{a_j^{(m)}, b_j^{(m)}\}$ at the mth step can be written as

$$\sum_{i,j} C_{kji} a_j^{(m)} a_i^{(m+1)} + \sum_{i,j} D_{kji} b_j^{(m)} a_i^{(m+1)} - \sum_i E_{kj} c^{(m+1)} + \sum_j K_{kj} a_j^{(m+1)}$$

$$= \int_{\partial G_1} r_1 \hat{\varphi}_k \, d\sigma + \iint_G \rho f_1 \hat{\varphi}_k \, dx \, dy \qquad k = 1, 2, \ldots \tag{13.3.4}$$

$$\sum_{i,j} C_{kji} a_j^{(m)} b_i^{(m+1)} + \sum_{i,j} D_{kji} b_j^{(m)} b_i^{(m+1)} - \sum_i F_{kj} c_j^{(m+1)} + \sum_j K_{kj} b_i^{(m+1)}$$

$$= \int_{\partial G_2} r_2 \hat{\varphi}_k \, d\sigma + \iint_G \rho f_2 \hat{\varphi}_k dx \, dy \qquad k = 1, 2, \ldots \tag{13.3.5}$$

$$\sum_j E_{jk} a_j^{(m+1)} + \sum_j F_{jk} b_i^{(m+1)} = 0 \qquad k = 1, 2, \ldots \tag{13.3.6}$$

where

$$C_{kji} = \iint_G \rho \hat{\varphi}_k \hat{\varphi}_j \frac{\partial \hat{\varphi}_i}{\partial x} \, dx \, dy \tag{13.3.7}$$

$$D_{kji} = \iint_G \rho \hat{\varphi}_k \hat{\varphi}_j \frac{\partial \hat{\varphi}_i}{\partial y} \, dx \, dy \tag{13.3.8}$$

$$E_{kj} = \iint_G \frac{\partial \hat{\varphi}_k}{\partial x} \hat{\psi}_j \, dx \, dy \tag{13.3.9}$$

$$F_{kj} = \iint_G \frac{\partial \hat{\varphi}_k}{\partial y} \hat{\psi}_j \, dx \, dy \tag{13.3.10}$$

$$K_{kj} = \iint_G \mu \left(\frac{\partial \hat{\varphi}_k}{\partial x} \frac{\partial \hat{\varphi}_j}{\partial x} + \frac{\partial \hat{\varphi}_k}{\partial y} \frac{\partial \hat{\varphi}_j}{\partial y} \right) dx \, dy \tag{13.3.11}$$

Thus, in using the iterative method for the present problem, starting from an appropriate initial guess $\{a_j^{(0)}, b_j^{(0)}, c_j^{(0)}\}$, we repeat the iteration given above until convergence is attained. By substituting the values converged into (13.3.1) through (13.3.3) we obtain the approximate solution sought.

13.4   *A Minimum Surface Problem*

In this section we consider a curve Γ given in three-dimensional xyz space. The problem is to find a surface with minimum area whose boundary is the curve Γ, and it is called the *minimum surface problem* or *Plateau's problem*.

For simplicity we assume that the boundary Γ is defined by a single-valued function

$$z = g(x, y) \tag{13.4.1}$$

The problem is to minimize the area

$$J[u] = \iint_G (1 + u_x^2 + u_y^2)^{1/2} \, dx \, dy \tag{13.4.2}$$

of the surface $z = u(x, y)$ surrounded by the boundary Γ, where G is the domain in the xy plane surrounded by the curve ∂G which is the projection of Γ onto the xy plane (Fig. 13.2). Based on the condition that the first variation of $J[u]$ vanishes, the Euler's equation for the present problem is

$$u_{xx}(1 + u_y^2) - 2u_{xy}u_x u_y + u_{yy}(1 + u_x^2) = 0 \tag{13.4.3}$$

Since the nonlinearity of this equation is strong, we cannot guarantee the convergence of an iterative method, as described in Sec. 13.1, in which a system of linear equations with respect to one unknown component chosen from each of $u_{xx}u_y^2$, $u_{xy}u_x u_y$, and $u_{yy}u_x^2$ is solved repeatedly.

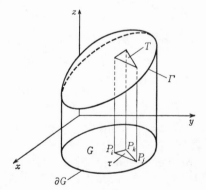

Fig. 13.2 The boundary Γ and the triangular element τ.

It is better in the present problem to use a method in which the functional $J[u]$ is minimized directly. Since only the first derivative of u appears in $J[u]$, we can use piecewise linear polynomials as the basis functions for a finite element approximation $\hat{u}_n$ to u. We again divide the domain G into triangular elements and construct a function $\hat{u}_n$ such that it is linear inside each triangular element τ and continuous in G. Let the three nodes of τ be P_i, P_j, and P_k, counterclockwise, and the values of $\hat{u}_n$ at each node by u_i, u_j, and u_k (Fig. 13.2). In other words, let the coordinates of the vertices of the triangle τ be (x_i, y_i, u_i), (x_j, y_j, u_j), and (x_k, y_k, u_k), which forms the part of the approximate polyhedron $z = \hat{u}_{n(x,y)}$ over the triangular element τ. Then $\hat{u}_n$ over τ can be expressed from (5.3.7) as

$$\hat{u}_n \Big|_\tau = \frac{1}{2S} \left\{ \begin{vmatrix} u_i & u_j & u_k \\ x_i & x_j & x_k \\ y_i & y_j & y_k \end{vmatrix} - \begin{vmatrix} u_i & u_j & u_k \\ 1 & 1 & 1 \\ y_i & y_j & y_k \end{vmatrix} x - \begin{vmatrix} u_i & u_j & u_k \\ x_i & x_j & x_k \\ 1 & 1 & 1 \end{vmatrix} y \right\}$$

$$(13.4.4)$$

where S is the area of the triangular element τ and, from (5.3.3), can be given by

$$S = \frac{1}{2} \begin{vmatrix} 1 & 1 & 1 \\ x_i & x_j & x_k \\ y_i & y_j & y_k \end{vmatrix} \qquad (13.4.5)$$

Therefore the total surface area of the polyhedron $z = \hat{u}_n(x, y)$ is

$$J[\hat{u}_n] = \sum_\tau S_T \qquad (13.4.6)$$

where S_T is the area of the triangle T over the triangular element τ and, from (13.4.4), is

$$S_T = \iint_\tau \left\{ 1 + \left(\frac{\partial \hat{u}_n}{\partial x} \right)^2 + \left(\frac{\partial \hat{u}_n}{\partial y} \right)^2 \right\}^{1/2} dx\, dy$$

$$= \frac{1}{2} \left\{ \begin{vmatrix} 1 & 1 & 1 \\ x_i & x_j & x_k \\ y_i & y_j & y_k \end{vmatrix}^2 + \begin{vmatrix} u_i & u_j & u_k \\ 1 & 1 & 1 \\ y_i & y_j & y_k \end{vmatrix}^2 + \begin{vmatrix} u_i & u_j & u_k \\ x_i & x_j & x_k \\ 1 & 1 & 1 \end{vmatrix}^2 \right\}^{1/2}$$

$$(13.4.7)$$

Thus our problem is reduced to a problem of finding a $\hat{u}_n$ that minimizes

$J[\hat{u}_n]$ in (13.4.6) under the condition that it satisfies

$$\hat{u}_n = g \qquad (13.4.8)$$

at the nodes on the boundary.

13.5 *A Minimum Surface Problem With a Multivalued Boundary Condition*

When we use the standard techniques of the FEM, we first divide the domain G into triangular elements and fix the nodes (x_i, y_i). Then we determine the values u_i of $\hat{u}_n$ at each node (x_i, y_i) so that $J[\hat{u}_n]$ in (13.4.6) becomes minimum. Note, however, that for a problem that minimizes $J[\hat{u}_n]$ in (13.4.6) variational parameters other than u_i may be used. That is, we can take not only u_i but also the coordinates x_i and y_i of the nodes as variational parameters according to the principle that the position of the point (x_i, y_i, u_i) in xyz space is determined so that $J[\hat{u}_n]$ is minimized. In this way the minimum of $J[\hat{u}_n]$ may become smaller and, in addition, we can formulate the variational form in a more flexible fashion. In fact, as seen in Chap. 12, we have obtained a good solution to a free boundary problem based on an idea similar to this. In this section we apply this idea to a minimum surface problem with a multivalued boundary condition and try to solve it.

As an example of a boundary on which a multivalued condition is given, we consider here a contour Γ consisting of two lines and two circular arcs as shown in Fig. 13.3. If the central angle θ is larger than $\pi/2$, the contour cannot be projected onto any plane in such a way that a boundary condition is given in a single-valued fashion on its projection ∂G. However, if we divide this contour into four equal parts based on its symmetry and restrict ourselves to the part shown in Fig. 13.4, the function u giving the minimum surface becomes a single-valued function of x

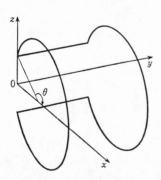

Fig. 13.3 The curve Γ for a minimum surface problem.

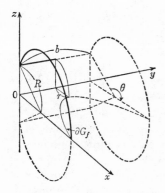

Fig. 13.4 Part of the domain whose boundary includes a free boundary.

and y, hence the boundary condition can be given in a single-valued fashion. Note that the cross section ∂G_f of the surface and the xy plane is unknown until the minimum surface is determined. This cross section can be treated as a kind of free boundary, as stated in Chap. 12. Thus the coordinates of the nodes along the free boundary ∂G_f can be taken as variational parameters, as mentioned in the previous section. On this free boundary the minimum surface $z = u(x,y)$ satisfies $\partial u/\partial x = \infty$; that is, $\partial x/\partial u = 0$ because of its symmetry with respect to the xy plane.

Let the length of the line segment of the contour be b, the radius of the circular arc be R, and the central angle be 2θ. Then, as noted, the boundary condition of the present problem can be given as

$$u = R \sin \theta \qquad \text{on } x = 0, \ 0 \le y \le \frac{b}{2} \qquad (13.5.1)$$

$$u = \{R^2 - (x + R \cos \theta)^2\}^{1/2} \qquad \text{on } y = 0, \ 0 \le x \le R(1 - \cos \theta) \qquad (13.5.2)$$

$$\frac{\partial u}{\partial y} = 0 \qquad \text{on } y = \frac{b}{2}, \ 0 \le x \le r \qquad (13.5.3)$$

$$u = 0 \qquad \frac{\partial x}{\partial u} = 0 \qquad \text{on the free boundary } \partial G_f \qquad (13.5.4)$$

The distance r from the y axis to the free boundary ∂G_f is unknown until the solution is obtained.

The domain G is partitioned as follows. First the angle θ of the circular arc on the xz plane at $y = 0$ is partitioned into M equal small angles $\Delta\theta$. The foot $(x_{i,0}, 0)$ of the perpendicular from each equally distributed partition point on the arc to the x axis is assigned as a node on the x axis, where

$$x_{i,0} = R\{1 - \cos(i \, \Delta\theta)\} \qquad \Delta\theta = \frac{\theta}{M} \quad i = 0, 1, \dots, M \qquad (13.5.5)$$

Next the interval $0 \leq y \leq b/2$ on the y axis is partitioned into N subintervals with mesh size Δy. Each equally distributed partition point $(0, y_j)$ is assigned as a node on the y axis, where

$$y_j = j \, \Delta y \qquad \Delta y = \frac{b}{2N} \qquad j = 0, 1, \ldots, N \qquad (13.5.6)$$

Then the cross-point $(x_{M,j}, y_j)$ of the line through each node $(0, y_j)$ on the y axis parallel to the x axis and the free boundary ∂G_j is assigned as a node on ∂G_j. Since $x_{M,j}$ is the x coordinate of the node on the free boundary, it is an unknown parameter. Finally the interval $0 \leq x \leq x_{M,j}$ at $y = y_j$ is partitioned into subintervals with an equal ratio of the distribution of $x_{i,0}$. Thus the coordinates of all the nodes are parameterized through unknowns $x_{M,j}$ along ∂G_f (Fig. 13.5). The x coordinate $x_{i,j}$ of each node at $(x_{i,j}, y_i)$ can be written as

$$x_{i,j} = \frac{x_{M,j} x_{i,0}}{x_{M,0}} \qquad (13.5.7)$$

The boundary condition (13.5.1) and (13.5.2) can be assigned as it is prescribed at each boundary node. To assign condition (13.5.3) we take (13.4.7) for the triangular element τ adjacent to the boundary and put the third term corresponding to $\partial \hat{u}_n / \partial y = 0$ inside the brace on the right-hand side of (13.4.7). Similarly, the second boundary condition of (13.5.4) can be assigned not by regarding the triangle T as a function u of (x, y) but by regarding it as a function x of (y, u). That is, if we express the triangle T adjacent to the free boundary as

$$x = x(y, u) \qquad (13.5.8)$$

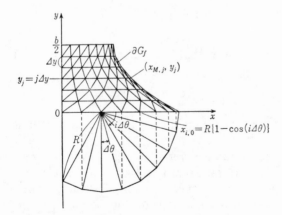

Fig. 13.5 Subdivision of the domain.

then its area can be given by

$$S_T = \iint \{1 + x_y^2 + x_u^2\}^{1/2} \, dy \, du \qquad (13.5.9)$$

This of course coincides with the right-hand side of (13.4.7). On the other hand, the derivative $\partial x/\partial n$ to which the second condition of (13.5.4) is assigned is given by

$$\frac{\partial x}{\partial u} = - \begin{vmatrix} 1 & 1 & 1 \\ x_i & x_j & x_k \\ y_i & y_j & y_k \end{vmatrix} \Bigg/ \begin{vmatrix} u_i & u_j & u_k \\ 1 & 1 & 1 \\ y_i & y_j & y_k \end{vmatrix} \qquad (13.5.10)$$

so that, in order to assign condition (13.5.4), we have only to make the first term zero inside the brace on the right-hand side of (13.4.7).

Among the variables $\{x_i, y_i, u_i\}$ appearng in $J[\hat{u}_n]$, which is going to be minimized, the variables of variation are the x coordinates $x_{M,j}$, $j = 1, 2, \ldots, N$, of the nodes on the free boundary and all the values u_i at the nodes for which boundary values are not prescribed. We rewrite all these unknown parameters as v_i, $i = 1, 2, \ldots, p$. Then our problem is reduced to minimization of the functional $J[v_1, v_2, \ldots, v_p]$ under the boundary condition (13.5.1) through (13.5.4), that is, to a problem of solving a system of nonlinear equations:

$$\frac{\partial J}{\partial v_i} \equiv f_i = 0 \qquad i = 1, 2, \ldots, p \qquad (13.5.11)$$

In general, although it is not always easy to solve such a system of nonlinear equations, we can make a good guess about the solution in the present problem, and in such a case the *generalized Newton method,* a kind of iterative method, is a good choice. First we guess an initial value $\{v_i^{(0)}\}$ as close as possible to $\{v_i\}$. Then we compute the $(m + 1)$th approximation $\{v_i^{(m+1)}\}$ from the mth approximation $\{v_i^{(m)}\}$ according to the iterative scheme

$$v_i^{(m+1)} = v_i^{(m)} - \omega \frac{f_i(v_1^{(m+1)}, \ldots, v_{i-1}^{(m+1)}, v_i^{(m)}, v_{i+1}^{(m)}, \ldots, v_p^{(m)})}{\partial f_i(v_1^{(m+1)}, \ldots, v_{i-1}^{(m+1)}, v_i^{(m)}, v_{i+1}^{(m)}, \ldots, v_p^{(m)})/\partial v_i} \qquad (13.5.12)$$

If the iteration converges, the limit will give the location of the nodes on the approximate minimum surface. Here ω is a parameter used to accelerate the convergence, and it is observed that in this problem convergence occurs quickly if ω is chosen close to 1.2.

As an example we show a finite element solution in Fig. 13.6 in which $R = 1.0$, $\theta = 5\pi/6$, $b = 1.5$, and the partition of the domain is such that $M = 15$ ($\Delta\theta = 6°$) and $N = 6$. In this example a *bifurcation* characteristic of nonlinear problems is observed; that is, the iteration converges to two

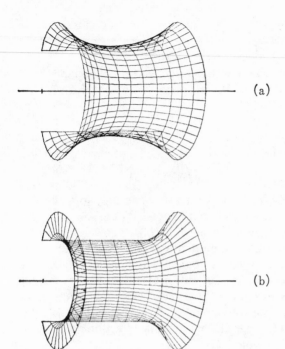

(a)

(b)

Fig. 13.6 Two solutions for $R = 1.0$, $\theta = 5\pi/6$, $b = 1.5$ (*From* M. Hinata, M. Shimasaki, T. Kiyono: Numerical solution of Plateau's problem by a finite element method, *Math. Comp.*, 28:45–60, 1974.

different solutions depending on the nature of the initial guess $\{v_i^{(0)}\}$, as shown in Fig. 13.6. There is a large difference between the r distances (from the y axis to the free boundary ∂G_f) for these two solutions, the result of computation being $r = 1.2606$ for Fig. 13.6A and $r = 0.2624$ for Fig. 13.6B. When $R = 1.0$ and $\theta = 5\pi/6$, this bifurcation is observed for b satisfying $1.4 < b < 1.53$, and for other values of b the solution is unique.

14

The Dual Variational Principle

14.1 *A Minimum Variational Problem*

The variational problems presented in this book so far in relation to the FEM were minimum variational problems in which a functional $J[u]$ was minimized. However, this type of minimum problem can usually be transformed into a maximum problem that is a dual problem corresponding to the original one. In this chapter, we discuss the transformation of a variational problem and present, as an application, a method of a posteriori error estimation of a finite element solution. We also discuss the treatment of the constraints on the problem.

Consider first the following minimum variational problem involving a functional $J[u]$ in a two-dimensional domain G.

Problem 1 (minimum variational problem)

Under the condition

$$u = g(x, y) \qquad \text{on } \partial G \tag{14.1.1}$$

minimize

$$J[u] = \iint_G F(u, u_x, u_y) \, dx \, dy \tag{14.1.2}$$

Here $F(u, u_x, u_y)$ is a function of u, u_x, and u_y, and

$$u_x = \frac{\partial u}{\partial x}$$

$$\text{(14.1.3)}$$

$$u_y = \frac{\partial u}{\partial y}$$

For admissible functions u we require differentiability such that the right-hand side of (14.1.2) is properly defined in addition to (14.1.1). Although this analysis also applies to a more general $F(u, u_x, u_y)$ provided that it satisfies appropriate conditions and that $J[u]$ has a minimum solution, in this section we restrict ourselves to

$$F(u, u_x, u_y) = \tfrac{1}{2}(u_x^2 + u_y^2 + qu^2 - 2fu) \qquad \text{(14.1.4)}$$

which we have frequently discussed, for example, in Chap. 4. In this case the admissible function must belong to H_1 as defined in Sec. 4.1.

We assume that the minimum variational problem (14.1.1) and (14.1.2) has a solution u^0; that is, we assume that, under the condition that u satisfies $u = g(x, y)$ on ∂G,

$$\min_u J[u] = J[u^0] = Q \qquad \text{(14.1.5)}$$

holds. In this chapter we attach the superscript 0 to the exact solution of the variational problem under consideration. Here Q is the minimum of $J[u]$, so that in general

$$Q \leq J[u] \qquad \text{(14.1.6)}$$

holds. In fact, according to the formulation stated in Sec. 4.2, it is evident that, if $q \geq 0$, then there is a solution of Prob. 1 when $F(u, u_x, u_y)$ is given by (14.1.4).

As already seen, the solution u^0 of the minimum problem makes the first variation of J vanish; that is, if we write any variation as $v = \delta u$, which satisfies $\delta u = 0$ on ∂G, then u^0 is given by a solution of the equation in the weak form

$$\delta J = \iint_G \left(F_u v + F_{u_x} \frac{\partial v}{\partial x} + F_{u_y} \frac{\partial v}{\partial y} \right) dx\, dy = 0 \qquad \forall v \in \overset{\circ}{H}_1$$

$$\text{(14.1.7)}$$

$$u = g \qquad \text{on } \partial G \qquad \text{(14.1.8)}$$

where F_u, F_{u_x}, and F_{u_y} are the partial derivatives of F with respect to u, u_x, and u_y, respectively, F being regarded as a function of x, y, u, u_x, and u_y. Hereafter we assume, together with the differentiability of F mentioned above, that F_u, F_{u_x}, and F_{u_y} belong to H_1. Here $\overset{\circ}{H}_1$ is a subspace of H_1 satisfying (4.1.7) as defined in Sec. 4.1. When F is given by (14.1.4), (14.1.7) coincides with (14.1.13).

If F_{u_x} and F_{u_y} have first-order partial derivatives with respect to x and

y, we have, integrating (14.1.7) by parts,

$$\delta J = \iint_G \left(F_u - \frac{\partial}{\partial x} F_{u_x} - \frac{\partial}{\partial y} F_{u_y} \right) v \, dx \, dy$$

$$+ \int_{\partial G} \left(F_{u_x} \frac{\partial x}{\partial n} + F_{u_y} \frac{\partial y}{\partial n} \right) v \, d\sigma = 0 \qquad (14.1.9)$$

Since $v = 0$ on ∂G, the second term vanishes and we obtain the following boundary value problem corresponding to the Euler's equation for the variational problem given above:

$$F_u - \frac{\partial}{\partial x} F_{u_x} - \frac{\partial}{\partial y} F_{u_y} = 0 \qquad (14.1.10)$$

$$u = g \qquad \text{on } \partial G \qquad (14.1.11)$$

In the case of example (14.1.4), (14.1.10) is just (4.1.1).

14.2 *An Additional Constraint Condition*

Although u_x and u_y represent the derivatives of u as defined by (14.1.3), in Prob. 1 we regard them as two new dependent variables related to u through the constraint conditions

$$u_x - \frac{\partial u}{\partial x} = 0 \qquad (14.2.1)$$

$$u_y - \frac{\partial u}{\partial y} = 0 \qquad (14.2.2)$$

Then we apply the *Lagrange method of multipliers* in order to handle the constraints in the variation. Hence we introduce two new functions $\lambda_1(x,y)$ and $\lambda_2(x,y)$ corresponding to the constraints (14.2.1) and (14.2.2) and define a functional H_λ:

$$H_\lambda[u, u_x, u_y] = \iint_G \left\{ F(u, u_x, u_y) + \lambda_1(x,y)\left(\frac{\partial u}{\partial x} - u_x\right) \right.$$

$$\left. + \lambda_2(x,y)\left(\frac{\partial u}{\partial y} - u_y\right) \right\} dx \, dy \qquad (14.2.3)$$

In the Lagrange method of multipliers λ_1 and λ_2 are usually simple numbers, but in the present case they are both functions of x and y. Furthermore, suppose that λ_1 and λ_2 have first-order derivatives in G; that is, they belong to H_1. Integrating by parts using the identity

$$\iint_G \frac{\partial}{\partial x} (\lambda_1 u) \, dx \, dy = \int_{\partial G} \lambda_1 u \frac{\partial x}{\partial n} \, d\sigma \qquad (14.2.4)$$

and a similar identity with respect to y, we have from (14.2.3)

$$H_\lambda[u, u_x, u_y] = \iint_G \left\{ F - \left(\frac{\partial\lambda_1}{\partial x} + \frac{\partial\lambda_2}{\partial y} \right) u - \lambda_1 u_x - \lambda_2 u_y \right\} dx\, dy$$

$$+ \int_{\partial G} \left(\lambda_1 \frac{\partial x}{\partial n} + \lambda_2 \frac{\partial y}{\partial n} \right) g\, d\sigma \qquad (14.2.5)$$

Here we have used condition (14.1.1).

If we are trying to make the new functional H_λ stationary in accordance with the Lagrange method of multipliers in the strict sense, we should also vary λ_1 and λ_2 together with u, u_x, and u_y. However, in order to understand how to obtain the maximum variational problem, it will be better to regard λ_1 and λ_2 as fixed functions when making H_λ stationary. Hence we consider the following intermediate problem.

Problem 2 (intermediate variational problem)

Make $H_\lambda[u, u_x, u_y]$ stationary under condition (14.1.1) where $\lambda_1(x, y)$ and $\lambda_2(x, y)$ are fixed functions belonging to H_1.

Problem 1 is a minimum problem involving $J[u]$, and the functional H_λ in the present problem consists of J plus linear terms with respect to u, u_x, and u_y. Therefore Prob. 2 is also a minimum problem. Let the solution of the minimum problem be $(u^\lambda, u_x^\lambda, u_y^\lambda)$, that is,

$$\min_{u, u_x, u_y} H_\lambda[u, u_x, u_y] = H_\lambda[u^\lambda, u_x^\lambda, u_y^\lambda] = Q_\lambda \qquad (14.2.6)$$

where Q_λ is the minimum.

In order to obtain the solution that attains the minimum of Q_λ we find the first variation δH_λ of $H_\lambda[u, u_x, u_y]$ in (14.2.5):

$$\delta H_\lambda = \iint_G \left[\left\{ F_u - \left(\frac{\partial\lambda_1}{\partial x} + \frac{\partial\lambda_2}{\partial y} \right) \right\} \delta u + \{ F_{u_x} - \lambda_1 \} \delta u_x \right.$$

$$\left. + \{ F_{u_y} - \lambda_2 \} \delta u_y \right] dx\, dy \qquad (14.2.7)$$

Then, from the minimum condition on H_λ, that is,

$$\delta H_\lambda = 0 \qquad (14.2.8)$$

we have the following three equations:

$$F_u = \frac{\partial\lambda_1}{\partial x} + \frac{\partial\lambda_2}{\partial y} \qquad (14.2.9)$$

$$F_{u_x} = \lambda_1 \qquad (14.2.10)$$

$$F_{u_y} = \lambda_2 \qquad (14.2.11)$$

If λ_1 and λ_2 are given functions, the set of functions (u, u_x, u_y) satisfying (14.1.1) together with (14.2.9) through (14.2.11) attains the minimum Q_λ of H_λ.

If we compare the minimum Q_1 with the minimum Q of the original Prob. 1, we see that

$$Q_\lambda \leq Q \qquad (14.2.12)$$

because, if we add new constraints $u_x - \partial u/\partial x = 0$ and $u_y - \partial u/\partial y = 0$ to Prob. 2 it is reduced to Prob. 1, while if we add new constraints to a minimum problem the minimum generally becomes larger.

14.3   The Lagrange Method of Multipliers

Now we return to the original Lagrange method of multipliers. We vary λ_1 and λ_2 together with u, u_x, and u_y, which we tentatively fixed in Prob. 2. To represent the present problem explicitly we write H, instead of (14.2.5), as

$$H[u, u_x, u_y, \lambda_1, \lambda_2] = \iint_G \left\{ F - \left(\frac{\partial \lambda_1}{\partial x} + \frac{\partial \lambda_2}{\partial y} \right) u - \lambda_1 u_x - \lambda_2 u_y \right\} dx\, dy$$

$$+ \int_{\partial G} \left(\lambda_1 \frac{\partial x}{\partial n} + \lambda_2 \frac{\partial y}{\partial n} \right) g\, d\sigma \qquad (14.3.1)$$

Then the first variation δH of $H[u, u_x, u_y, \lambda_1, \lambda_2]$ becomes

$$\delta H = \iint_G \left[\left\{ F_u - \left(\frac{\partial \lambda_1}{\partial x} + \frac{\partial \lambda_2}{\partial y} \right) \right\} \delta u + \{ F_{u_x} - \lambda_1 \} \delta u_x + \{ F_{u_v} - \lambda_2 \} \delta u_y \right.$$

$$\left. - \left(\frac{\partial \delta\lambda_1}{\partial x} + \frac{\partial \delta\lambda_2}{\partial y} \right) u - u_x\, \delta\lambda_1 - u_y\, \delta\lambda_2 \right] dx\, dy$$

$$+ \int_{\partial G} \left(\delta\lambda_1 \frac{\partial x}{\partial n} + \delta\lambda_2 \frac{\partial y}{\partial n} \right) g\, d\sigma$$

$$= \iint_G \left[\left\{ F_u - \left(\frac{\partial \lambda_1}{\partial x} + \frac{\partial \lambda_2}{\partial y} \right) \right\} \delta u + (F_{u_x} - \lambda_1)\, \delta u_x + (F_{u_y} - \lambda_2)\, \delta u_y \right.$$

$$\left. + \left(\frac{\partial u}{\partial x} - u_x \right) \delta\lambda_1 + \left(\frac{\partial u}{\partial y} - u_y \right) \delta\lambda_2 \right] dx\, dy$$

$$+ \int_{\partial G} \left(\delta\lambda_1 \frac{\partial x}{\partial n} + \delta\lambda_2 \frac{\partial y}{\partial n} \right) (g - u)\, d\sigma \qquad (14.3.2)$$

Therefore, from the stationary condition of H

$$\delta H = 0 \qquad (14.3.3)$$

we obtain the following six equations:

$$F_u = \frac{\partial \lambda_1}{\partial x} + \frac{\partial \lambda_2}{\partial y} \tag{14.3.4}$$

$$F_{u_x} = \lambda_1 \tag{14.3.5}$$

$$F_{u_y} = \lambda_2 \tag{14.3.6}$$

$$\frac{\partial u}{\partial x} = u_x \tag{14.3.7}$$

$$\frac{\partial u}{\partial y} = u_y \tag{14.3.8}$$

$$u = g \quad \text{on } \partial G \tag{14.3.9}$$

As seen from this result, $\delta u = 0$ need not be imposed on the variation δu of u on the boundary ∂G when we find the first variation of H, because the variation of H leads to (14.3.9) as a natural condition. The reason for this is that we use (14.1.1) when we define (14.2.5) from (14.2.3). Also, the fact that condition (14.3.7) and (14.3.8) is obtained as a natural condition is a natural result of the Lagrange method of multipliers.

Let the exact solution of (14.3.4) through (14.3.9) be $\{u^0, u_x^0, u_y^0, \lambda_1^0, \lambda_2^0\}$. Take any function v in $\mathring{H}_1$ satisfying $v = 0$ on ∂G and consider the integral

$$\iint_G \left\{ \left(\frac{\partial \lambda_1^0}{\partial x} + \frac{\partial \lambda_2^0}{\partial y} \right) v + \lambda_1^0 \frac{\partial v}{\partial x} + \lambda_2^0 \frac{\partial v}{\partial y} \right\} dx\, dy \tag{14.3.10}$$

Integrate it by parts taking into account the condition that $v = 0$ on ∂G. Then we see that the integral (14.3.10) vanishes. Therefore, from (14.3.4) through (14.3.6) the solution $u = u^0$ satisfies

$$\iint_G \left(F_u v + F_{u_x} \frac{\partial v}{\partial x} + F_{u_y} \frac{\partial v}{\partial y} \right) dx\, dy = 0 \quad \forall v \in \mathring{H}_1 \tag{14.3.11}$$

and, from (14.3.9), u^0 coincides with g on ∂G.

By comparing (14.3.11) and (14.1.7) we conclude that the solution u^0 obtained from $\delta H = 0$ is identical to the solution of Prob. 1. Since the Lagrange method of multipliers is formulated so as to obtain this result, this is a natural consequence. Furthermore, we see that, from definition (14.2.3) to which the functional H is ascribed, the stationary value of H given by (14.3.4) through (14.3.9) is equal to the minimum Q of J.

14.4 A Dual Variational Problem

In Sec. 14.2 we saw that, if u, u_x, and u_y satisfy (14.1.1) and (14.2.9) through (14.2.11) or satisfy (14.1.1) and (14.3.4) through (14.3.6) for given functions λ_1, λ_2, and H_λ, the minimum Q_λ of H_λ satisfies

$$Q_\lambda \leq Q \tag{14.4.1}$$

Also, we saw in the previous section that Q_λ coincides with Q if (14.3.7) and (14.3.8) hold for u, u_x, and u_y, which satisfy the same condition mentioned above by an appropriate choice of λ_1 and λ_2. Thus we obtain the following important relation:

$$\max_{\lambda_1,\lambda_2 \in H_1} Q_\lambda = Q \tag{14.4.2}$$

Consequently, based on the problem in which the functional $H[u,u_x,u_y,\lambda_1,\lambda_2]$ is made stationary, the following new problem is obtained. That is, by regarding (14.3.4) through (14.3.6) as constraints on λ_1 and λ_2 through u, u_x, and u_y, we vary λ_1 and λ_2 and search for the stationary point of H under these constraints. From (14.4.2) this problem is apparently a maximum problem.

In order to write this maximum problem explicitly we need to express u, u_x, and u_y as functions of x, y, λ_1, and λ_2 from (14.3.4) through (14.3.6). We examine this situation in example (14.1.4), and in this case we have

$$F_u = qu - f \tag{14.4.3}$$

$$F_{u_x} = u_x \tag{14.4.4}$$

$$F_{u_y} = u_y \tag{14.4.5}$$

hence we immediately see that (14.3.5) and (14.3.6) become

$$u_x = \lambda_1 \tag{14.4.6}$$

$$u_y = \lambda_2 \tag{14.4.7}$$

Now we assume that $q > 0$ for all (x,y) in the domain. Then (14.4.3) can be solved for u, and condition (14.3.4) becomes

$$u = \frac{1}{q}\left(\frac{\partial \lambda_1}{\partial x} + \frac{\partial \lambda_2}{\partial y} + f\right) \tag{14.4.8}$$

If we substitute (14.4.6) through (14.6.8) for F in (14.1.4) and substitute them into the integral over G in the functional H in (14.3.1), it can be expressed in terms of a function that includes only λ_1 and λ_2:

$$F - \left(\frac{\partial \lambda_1}{\partial x} + \frac{\partial \lambda_2}{\partial y}\right)u - \lambda_1 u_x - \lambda_2 u_y$$

$$= -\frac{1}{2}\left\{\lambda_1^2 + \lambda_2^2 + \frac{1}{q}\left(\frac{\partial \lambda_1}{\partial x} + \frac{\partial \lambda_2}{\partial y} + f\right)^2\right\} \tag{14.4.9}$$

Thus we obtain a new variational problem involving the maximization of H_λ in (14.2.5) or H in (14.3.1), varying λ_1 and λ_2.

The procedure stated above can be summarized in a more general form as follows. Suppose that the integrand of the integral over G in (14.3.1) is expressed using (14.3.4) through (14.3.6) in terms of a function $-\Psi(\lambda_1, \lambda_2)$ that includes only λ_1 and λ_2:

$$-\Psi(\lambda_1, \lambda_2) = F - \left(\frac{\partial \lambda_1}{\partial x} + \frac{\partial \lambda_2}{\partial y}\right)u - \lambda_1 u_x - \lambda_2 u_y \quad (14.4.10)$$

where λ_1 and λ_2 are assumed to be so smooth that (14.4.10) is integrable over G. In example (14.4.9) they have only to belong to H_1.

Problem 3 (maximum variational problem)

Maximize

$$I[\lambda_1, \lambda_2] = -\iint_G \Psi(\lambda_1, \lambda_2) \, dx \, dy + \int_{\partial G} \left(\lambda_1 \frac{\partial x}{\partial n} + \lambda_2 \frac{\partial y}{\partial n}\right) g \, d\sigma \quad (14.4.11)$$

There is no constraint on λ_1 and λ_2 in this variational problem. Note that constraint (14.1.1) in the original minimum variational problem is taken into account when (14.3.1) is derived using integration of H by parts.

While the minimum of $J[u]$ in Prob. 1 is Q, the maximum of $I[\lambda_1, \lambda_2]$ in Prob. 3 is also Q. Therefore, for any u having appropriate differentiability and satisfying (14.1.1), and for any $\lambda_1, \lambda_2 \in H_1$, the inequality

$$I[\lambda_1, \lambda_2] \leq J[u] \quad (14.4.12)$$

holds.

The transformation defined by (14.3.4) through (14.3.6) and (14.4.10) for the original minimum variational Prob. 1 is called a *Friedrichs transformation* or a *contact transformation*. This transformation is also a kind of *Legendre transformation*. The transformed variational Prob. 3 is called a *dual variational problem* corresponding to the original Prob. 1. A dual variational problem is sometimes called a *reciprocal variational problem* or a *complementary variational problem*. It is easy to see that Prob. 1 is derived from Prob. 3 by applying a Friedrichs transformation to the latter. Thus an identical problem is formulated in dual forms, the minimization of J and the maximization of I.

We derive here the weak form equation of the dual problem corresponding to problem (14.1.4). The first variation of the functional $I[\lambda_1, \lambda_2]$ is equal to the first variation of the functional $H[u, u_x, u_y, \lambda_1, \lambda_2]$ under constraints (14.3.4) through (14.3.6). Therefore, from the first equation in (14.3.2) and (14.4.6) through (14.4.8), the first variation δI of I becomes

$$\delta I = -\iint_G \left\{ \left(\frac{\partial \delta \lambda_1}{\partial x} + \frac{\partial \delta \lambda_2}{\partial y} \right) u + u_x \, \delta \lambda_1 + u_y \, \delta \lambda_2 \right\} dx \, dy$$

$$+ \int_{\partial G} \left(\delta \lambda_1 \frac{\partial x}{\partial n} + \delta \lambda_2 \frac{\partial y}{\partial n} \right) g \, d\sigma$$

$$= -\iint_G \left\{ \frac{1}{q} \left(\frac{\partial \lambda_1}{\partial x} + \frac{\partial \lambda_2}{\partial y} + f \right) \left(\frac{\partial \delta \lambda_1}{\partial x} + \frac{\partial \delta \lambda_2}{\partial y} \right) + \lambda_1 \, \delta \lambda_1 + \lambda_2 \, \delta \lambda_2 \right\} dx \, dy$$

$$+ \int_{\partial G} \left(\delta \lambda_1 \frac{\partial x}{\partial n} + \delta \lambda_2 \frac{\partial y}{\partial n} \right) g \, d\sigma$$

$$(14.4.13)$$

If we use μ_1 and μ_2 instead of $\delta \lambda_1$ and $\delta \lambda_2$, we obtain from $\delta I = 0$ the following equation in the weak form corresponding to the dual variational problem (14.1.4):

$$\iint_G \left\{ \frac{1}{q} \left(\frac{\partial \lambda_1}{\partial x} + \frac{\partial \lambda_2}{\partial y} + f \right) \left(\frac{\partial \mu_1}{\partial x} + \frac{\partial \mu_2}{\partial y} \right) + \lambda_1 \mu_1 + \lambda_2 \mu_2 \right\} dx \, dy$$

$$= \int_{\partial G} \left(\mu_1 \frac{\partial x}{\partial n} + \mu_2 \frac{\partial y}{\partial n} \right) g \, d\sigma \qquad \forall \mu_1, \mu_2 \in H_1$$

$$(14.4.14)$$

Furthermore, from the second equation of (14.3.2) or from the integration of (14.4.14) by parts, we obtain the following Euler's equation.

Under the constraint that

$$\frac{1}{q} \left(\frac{\partial \lambda_1}{\partial x} + \frac{\partial \lambda_2}{\partial y} + f \right) = g \qquad \text{on } \partial G \qquad (14.4.15)$$

solve

$$-\frac{\partial}{\partial x} \left\{ \frac{1}{q} \left(\frac{\partial \lambda_1}{\partial x} + \frac{\partial \lambda_2}{\partial y} + f \right) \right\} + \lambda_1 = 0 \qquad (14.4.16)$$

$$-\frac{\partial}{\partial y} \left\{ \frac{1}{q} \left(\frac{\partial \lambda_1}{\partial x} + \frac{\partial \lambda_2}{\partial y} + f \right) \right\} + \lambda_2 = 0 \qquad (14.4.17)$$

From (14.4.8) we see that (14.4.15) through (14.4.17) are equivalent to (4.1.1) and (4.1.2).

14.5 A Dual Variational Problem With a Constraint

In the previous section we assumed that (14.3.4) through (14.3.6) could be solved for u, u_x, and u_y as functions of λ_1 and λ_2 when we derived the dual variational problem. Actually it is not always possible to solve

(14.3.4) through (14.3.6) for u, u_x, and u_y. In fact, in example (14.1.4), $F_u = -f$ when $q \equiv 0$, so that (14.3.4), that is,

$$\frac{\partial \lambda_1}{\partial x} + \frac{\partial \lambda_2}{\partial y} = -f \qquad (14.5.1)$$

cannot be solved for u. However, (14.5.1) does not include u, u_x, and u_y, hence it is possible to regard it as a constraint on the variational problem in which λ_1 and λ_2 are varied. To show how to handle (14.5.1) as a constraint, we consider the following minimum variational problem involving a two-dimensional domain G.

Problem 4 (minimum variational problem)

Under the constraint that

$$u = g \qquad \text{on } \partial G \qquad (14.5.2)$$

minimize

$$J[u] = \frac{1}{2} \iint_G \left\{ \left(\frac{\partial u}{\partial x} \right)^2 + \left(\frac{\partial u}{\partial y} \right)^2 - 2fu \right\} dx\, dy \qquad (14.5.3)$$

for uH_1.

If we introduce a vector function $\boldsymbol{\lambda}$ whose components are λ_1 and λ_2

$$\boldsymbol{\lambda} = (\lambda_1, \lambda_2) \qquad (14.5.4)$$

we can write (14.5.1) as

$$\operatorname{div} \boldsymbol{\lambda} = -f \qquad (14.5.5)$$

Also, if we take into account (14.5.1) when $q \equiv 0$, (14.4.10) becomes

$$-\Psi(\lambda_1, \lambda_2) = -\tfrac{1}{2}(\lambda_1^2 + \lambda_2^2) = -\tfrac{1}{2}\boldsymbol{\lambda} \cdot \boldsymbol{\lambda} \qquad (14.5.6)$$

Thus we obtain the dual variational problem corresponding to Prob. 4.

Problem 5 (maximum variational problem with a constraint)

Under the constraint that

$$\operatorname{div} \boldsymbol{\lambda} = -f \qquad \boldsymbol{\lambda} = (\lambda_1, \lambda_2) \qquad \text{in } G \qquad (14.5.7)$$

maximize

$$I[\lambda_1, \lambda_2] = -\frac{1}{2} \iint_G (\lambda_1^2 + \lambda_2^2)\, dx\, dy + \int_{\partial G} \left(\lambda_1 \frac{\partial x}{\partial n} + \lambda_2 \frac{\partial y}{\partial n} \right) g\, d\sigma \qquad (14.5.8)$$

for $\lambda_1, \lambda_2 \in H_1$.

It will be evident in this case also that the inequality (14.4.12) holds for any $u \in H_1$ satisfying (14.5.2) and for any $\lambda_1, \lambda_2 \in H_1$ satisfying (14.5.7).

The equation in the weak form corresponding to Prob. 5 can be obtained from the first variation of (14.5.8) or directly from the first equation of (14.3.2) using (14.4.6) and (14.4.7). In the present case, however, the test function $(\lambda_1 + \delta\lambda_1, \lambda_2 + \delta\lambda_2)$ must satisfy (14.5.5), so $\delta\lambda_1$ and $\delta\lambda_2$ must satisfy

$$\frac{\partial \delta\lambda_1}{\partial x} + \frac{\partial \delta\lambda_2}{\partial y} = 0 \qquad (14.5.9)$$

Therefore, if we use μ_1 and μ_2 instead of $\delta\lambda_1$ and $\delta\lambda_2$ and write

$$\mu = (\mu_1, \mu_2) \qquad (14.5.10)$$

we obtain the following equation in the weak form from the first equation of (14.3.2).

Under the constraint that

$$\operatorname{div} \lambda = -f \qquad \text{in } G \qquad (14.5.11)$$

solve

$$-\iint_G (\lambda_1\mu_1 + \lambda_2\mu_2) \, dx \, dy + \int_{\partial G} \left(\mu_1 \frac{\partial x}{\partial n} + \mu_2 \frac{\partial y}{\partial n} \right) g \, d\sigma = 0$$
$$(14.5.12)$$

where μ_1 and μ_2 are arbitrary functions such that $\mu_1, \mu_2 \in H_1$ and satisfy

$$\operatorname{div} \mu = 0 \qquad \text{in } G \qquad (14.5.13)$$

Furthermore, integrating (14.5.2) by parts, we obtain the Euler's equation as follows.

Under the constraint that

$$\operatorname{div} \lambda = -f \qquad \text{in } G \qquad (14.5.14)$$

solve

$$\frac{\partial u}{\partial x} = \lambda_1 \qquad (14.5.15)$$

$$\frac{\partial u}{\partial y} = \lambda_2 \qquad (14.5.16)$$

$$u = g \qquad \text{on } \partial G \qquad (14.5.17)$$

Here u is regarded as a parameter through which λ_1 and λ_2 are determined. It is evident that the set of equations (14.5.14) through (14.5.17) is essentially identical to (4.1.1) and (4.1.2).

14.6 *Tangent Transformation*

To understand the dual variational principle in a more intuitive way we examine it from a geometric point of view. For simplicity we consider problem (14.5.3) with $f \equiv 0$. That is, the integrand of $J[u]$ is

$$F(u_x, u_y) = \tfrac{1}{2}(u_x^2 + u_y^2) \tag{14.6.1}$$

Writing u_x and u_y as ξ and η, respectively, we consider a surface

$$\zeta = F(\xi, \eta) = \tfrac{1}{2}(\xi^2 + \eta^2) \tag{14.6.2}$$

in three-dimensional $\xi\eta\zeta$ space corresponding to (14.6.1). This is an elliptic paraboloid of revolution that is differentiable and convex toward the bottom. The equation of the tangent plane at a point (ξ_0, η_0, ζ_0) on this surface (Fig. 14.1) is given by

$$\zeta = \lambda_1 \xi + \lambda_2 \eta - \Psi(\lambda_1, \lambda_2) \tag{14.6.3}$$

where λ_1, λ_2, and Ψ can be determined from the following three conditions which indicate that (14.6.3) is tangent to (14.6.2) at (ξ_0, η_0, ζ_0):

$$F(\xi_0, \eta_0) = \lambda_1 \xi_0 + \lambda_2 \eta_0 - \Psi(\lambda_1, \lambda_2) \tag{14.6.4}$$

$$F_{\xi_0} = \lambda_1 \tag{14.6.5}$$

$$F_{\eta_0} = \lambda_2 \tag{14.6.6}$$

This is just a Friedrichs transformation in which the function F includes u_x and u_y but does not include u explicitly. The third parameter Ψ in (14.6.3) is a function of λ_1 and λ_2 because only two of the λ_1, λ_2, and Ψ components can be varied independently since the degrees of freedom in the choice of the coordinates (ξ_0, η_0) is 2.

At the point of contact (ξ_0, η_0, ζ_0) between the surface (14.6.2) and the plane (14.6.3), the equality (14.6.4) holds. However, in general, at an arbi-

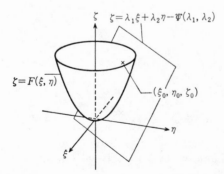

Fig. 14.1 The surface $\zeta = F(\xi, \eta)$ and its tangent plane.

trary point (ξ, η) the inequality

$$F(\xi, \eta) \geq \lambda_1 \xi + \lambda_2 \eta_2 - \Psi(\lambda_1, \lambda_2) \qquad (14.6.7)$$

holds because the tangent plane of a surface that is convex toward the bottom is always located below the surface. Thus, rewriting ξ and η as u_x and u_y, we have

$$F(u_x, u_y) - \lambda_1 u_x - \lambda_2 u_y \geq -\Psi(\lambda_1, \lambda_2) \qquad (14.6.8)$$

The point (ξ_0, η_0, ζ_0) of contact can be chosen arbitrarily on the surface (14.6.2), and the inequality (14.6.7) holds for an arbitrary point (ξ, η) on the $\xi\eta$ plane. On the other hand, the fact that the coordinates (ξ_0, η_0) of the point of contact can be chosen arbitrarily implies that λ_1 and λ_2 also can be chosen arbitrarily through the Friedrichs transformation (14.6.4) through (14.6.6). This means that the inequality (14.6.8) holds for any u_x, u_y, λ_1, and λ_2. The equality sign holds only for such λ_1 and λ_2 that correspond exactly to the situation $u_x = \xi_0$ and $u_y = \eta_0$ through (14.6.5) and (14.6.6).

If we integrate both sides of the inequality (14.6.8) over G, add to it the boundary integral

$$\int_{\partial G} \left(\lambda_1 \frac{\partial x}{\partial n} + \lambda_2 \frac{\partial y}{\partial n} \right) g \, d\sigma \qquad (14.6.9)$$

and integrate by parts using

$$u_x = \frac{\partial u}{\partial x} \qquad u_y = \frac{\partial u}{\partial y} \qquad (14.6.10)$$

and (14.2.4), we obtain

$$\frac{1}{2} \iint_G \left\{ \left(\frac{\partial u}{\partial x} \right)^2 + \left(\frac{\partial u}{\partial y} \right)^2 \right\} dx \, dy + \iint_G \left(\frac{\partial \lambda_1}{\partial x} + \frac{\partial \lambda_2}{\partial y} \right) u \, dx \, dy$$

$$- \int_{\partial G} \left(\lambda_1 \frac{\partial x}{\partial n} + \lambda_2 \frac{\partial y}{\partial n} \right) (u - g) \, d\sigma \geq -\frac{1}{2} \iint_G (\lambda_1^2 + \lambda_2^2) \, dx \, dy$$

$$+ \int_{\partial G} \left(\lambda_1 \frac{\partial x}{\partial n} + \lambda_2 \frac{\partial y}{\partial n} \right) g \, d\sigma \qquad (14.6.11)$$

Then, taking into account constraints (14.5.2) and (14.5.7), we eventually obtain the inequality

$$I[\lambda_1, \lambda_2] \leq J[u] \qquad (14.6.12)$$

where $u \in H_1$ is any function satisfying (14.5.2) and $\lambda_1, \lambda_2 \in H_1$ are any functions satisfying (14.5.7). This inequality is just (14.4.12).

The reason why the Friedrichs transformation is called the *tangent transformation* is clear from the analysis described above.

14.7 Estimation of the Upper and Lower Bounds of the Minimum of $J[u]$

A functional $J[u]$ encountered in an actual physical problem usually corresponds to some physical quantity, and its minimum Q is identical to a value of the physical quantity actually observed. A typical example is given in Sec. 14.9. In the dual variational problem, on the other hand, Q is also the maximum of the functional $I[\lambda_1, \lambda_2]$. Therefore, for any admissible function $\hat{u}$ and for any admissible function $\hat{\lambda}_1$ and $\hat{\lambda}_2$, the inequality

$$I[\hat{\lambda}_1, \hat{\lambda}_2] \leq Q \leq J[\hat{u}] \qquad (14.7.1)$$

holds. Thus, if we minimize $J[\hat{u}]$ by choosing an appropriate admissible function $\hat{u}$ and also, independently, maximize $I[\hat{\lambda}_1, \hat{\lambda}_2]$ by choosing an appropriate admissible function $\hat{\lambda}_1$ and $\hat{\lambda}_2$, the exact value Q of the physical quantity is estimated from above by an approximate value $J[\hat{u}]$ and also from below by an approximate value $I[\hat{\lambda}_1, \hat{\lambda}_2]$.

When solving a problem in practice, we first compute a finite element solution $\hat{u}$ of (14.1.7) and (14.1.8) in the weak form corresponding to the minimum problem and evaluate $J[\hat{u}]$ which serves as an upper bound of Q. Next we compute a finite element solution $(\hat{\lambda}_1, \hat{\lambda}_2)$ of the equation in the weak form corresponding to the dual variational problem, say (14.4.14), and evaluate $I[\hat{\lambda}_1, \hat{\lambda}_2]$ which serves as a lower bound of Q.

14.8 A Posteriori Error Estimation of the Finite Element Solution

When the functional $J[u]$ of the given minimum variational problem has an appropriate form, the error of a finite element solution $\hat{u}$ itself may be estimated. Again we consider $J[u]$ whose integrand is given by (14.1.4), that is,

$$J[u] = \frac{1}{2} \iint_G \left\{ \left(\frac{\partial u}{\partial x}\right)^2 + \left(\frac{\partial u}{\partial y}\right)^2 + qu^2 - 2fu \right\} dx\, dy \qquad (14.8.1)$$

The boundary condition is given as

$$u = g \qquad \text{on } \partial G \qquad (14.8.2)$$

First consider the case

$$q > 0 \qquad (14.8.3)$$

Here the dual variational problem corresponding to the minimum problem (14.8.1) is, from (14.4.9) through (14.4.11), equivalent to the maximum problem of the functional

$$I[\lambda_1, \lambda_2] = -\frac{1}{2} \iint_G \left\{ \lambda_1^2 + \lambda_2^2 + \frac{1}{q} \left(\frac{\partial \lambda_1}{\partial x} + \frac{\partial \lambda_2}{\partial y} + f \right)^2 \right\} dx \, dy$$

$$+ \int_{\partial G} \left(\lambda_1 \frac{\partial x}{\partial n} + \lambda_2 \frac{\partial y}{\partial n} \right) g \, d\sigma \qquad (14.8.4)$$

Let $\hat{u}$ be an arbitrary admissible function for the minimum variational problem (14.8.1) under constraint (14.8.2), and also let $\hat{\lambda}_1$ and $\hat{\lambda}_2$ be an arbitrary admissible function for the maximum variational problem (14.8.4). Define ϵ^2 by

$$\epsilon^2 = J[\hat{u}] - I[\hat{\lambda}_1, \hat{\lambda}_2] \qquad (14.8.5)$$

Note that ϵ^2 is computable if $\hat{u}$ and $\hat{\lambda}_1, \hat{\lambda}_2$ are obtained as finite element solutions of each variational problem. From (14.8.1) and (14.8.4) it is easy to see that ϵ^2 can be expressed as

$$\epsilon^2 = \frac{1}{2} \iint_G \left[\left(\frac{\partial \hat{u}}{\partial x} - \hat{\lambda}_1 \right)^2 + \left(\frac{\partial \hat{u}}{\partial y} - \hat{\lambda}_2 \right)^2 \right.$$

$$\left. + q \left\{ \hat{u} - \frac{1}{q} \left(\frac{\partial \hat{\lambda}_1}{\partial x} + \frac{\partial \hat{\lambda}_2}{\partial y} + f \right) \right\}^2 \right] dx \, dy \qquad (14.8.6)$$

Since $(\hat{\lambda}_1, \hat{\lambda}_2)$ is an arbitrary admissible function, we choose the exact solution $(\lambda_1^0, \lambda_2^0)$ for $(\hat{\lambda}_1, \hat{\lambda}_2)$ that maximizes $I[\lambda_1, \lambda_2]$ in (14.8.4). Then, from

$$I[\hat{\lambda}_1, \hat{\lambda}_2] \leq I[\lambda_1^0, \lambda_2^0] = Q \qquad (14.8.7)$$

we have

$$\epsilon^2 \geq J[\hat{u}] - Q$$

$$= \frac{1}{2} \iint_G \left[\left(\frac{\partial \hat{u}}{\partial x} - \lambda_1^0 \right)^2 + \left(\frac{\partial \hat{u}}{\partial y} - \lambda_2^0 \right)^2 \right.$$

$$\left. + q \left\{ \hat{u} - \frac{1}{q} \left(\frac{\partial \lambda_1^0}{\partial x} + \frac{\partial \lambda_2^0}{\partial y} + f \right) \right\}^2 \right] dx \, dy \qquad (14.8.8)$$

This inequality can be written, from (14.4.6) through (14.4.8), as

$$\epsilon^2 \geq \frac{1}{2} \iint_G \left[\left(\frac{\partial \hat{u}}{\partial x} - \frac{\partial u^0}{\partial x} \right)^2 + \left(\frac{\partial \hat{u}}{\partial y} - \frac{\partial u^0}{\partial y} \right)^2 + q \, (\hat{u} - u^0)^2 \right] dx \, dy \qquad (14.8.9)$$

The right-hand side is just the energy norm defined in Sec. 6.4. Thus the error of a finite element solution $\hat{u}$ can be estimated in terms of the energy norm using the computable quantity ϵ^2:

$$\| \hat{u} - u^0 \|_a \leq \epsilon \qquad (14.8.10)$$

This estimate gives an explicit upper bound for the error of $\hat{u}$ computed using the finite element solutions $\hat{u}$ and $\hat{\lambda}_1, \hat{\lambda}_2$ themselves after they are obtained. An estimate obtained in this way is called an *a posteriori error estimate*. On the other hand, an estimate obtained from an analysis using information from the given problem before solving it, as shown in Chaps. 3 and 6, is called an *a priori error estimate*. It is also possible, if necessary, to derive an expression of an a posteriori error estimate for a finite element solution $(\hat{\lambda}_1, \hat{\lambda}_2)$ of the dual variational problem in a similar way.

Once an estimate like (14.8.10) in terms of the energy norm is obtained, it is easy to express it in terms of the Sobolev norm by virtue of the ellipticity condition (4.3.8):

$$\|\hat{u} - u^0\|_1 \leq \frac{1}{\sqrt{\gamma}} \, \epsilon \qquad (14.8.11)$$

Next consider the case

$$q \equiv 0 \qquad (14.8.12)$$

Then, from (14.5.3) and (14.5.8)

$$\epsilon^2 = J[\hat{u}] - I[\hat{\lambda}_1, \hat{\lambda}_2]$$

$$= \frac{1}{2} \iint_G \left\{ \left(\frac{\partial \hat{u}}{\partial x} - \hat{\lambda}_1 \right)^2 + \left(\frac{\partial \hat{u}}{\partial y} - \hat{\lambda}_2 \right)^2 \right\} dx \, dy \qquad (14.8.13)$$

holds. By choosing the exact solution $(\lambda_1^0, \lambda_2^0)$ for $(\hat{\lambda}_1, \hat{\lambda}_2)$, we again have

$$\epsilon^2 \geq \frac{1}{2} \iint_G \left\{ \left(\frac{\partial \hat{u}}{\partial x} - \frac{\partial u^0}{\partial x} \right)^2 + \left(\frac{\partial \hat{u}}{\partial y} - \frac{\partial u^0}{\partial y} \right)^2 \right\} dx \, dy = \|\hat{u} - u^0\|_a^2$$

$$(14.8.14)$$

and, from the ellipticity condition (4.3.8), an estimate in terms of the Sobolev norm is obtained in the same form as (14.8.11).

In order to carry out a posteriori error estimation as described above, we have to solve two problems, the original problem and the dual problem. In cases in which the solution of the dual problem is significant, like a problem to be given in the next section, a posteriori error estimation may be done efficiently. Otherwise, however, it should be judged beforehand whether solving the dual problem is worthwhile or not. Furthermore, while the finite element solution $(\hat{\lambda}_1, \hat{\lambda}_2)$ must satisfy exactly the condition for an admissible function in the present error estimation, it is usually quite difficult for $(\hat{\lambda}_1, \hat{\lambda}_2)$ to satisfy it. For example, in the dual variational problem with $q \equiv 0$ the finite element solution must satisfy (14.5.7) in G, but it is difficult to accomplish this. On the other hand, when $q > 0$, it is fairly easy to solve the dual variational problem of maximizing

$I[\lambda_1, \lambda_2]$ in (14.4.11). Consequently, an a posteriori error estimation should be done only after sufficient consideration of its feasibility.

14.9 Physical Interpretation of the Dual Variational Principle

The dual variational principle is not only of mathematical interest but in many cases also has physical significance. We take here as an example the electrostatic capacity of a conductive material located in a dielectric material and examine the physical significance of the duality between u and λ_1, λ_2.

Suppose the shape of the conductive material and the dielectric constant do not change in a certain direction in three-dimensional space and that a constant electric charge is present on the surface of the conductive material. Then the *electrostatic potential u* in the exterior domain of the conductive material can be given by a solution of the following problem in a two-dimensional domain G in the plane perpendicular to the direction of the constant shape:

$$\text{div}(\epsilon \text{ grad } u) = \frac{\partial}{\partial x}\left(\epsilon \frac{\partial u}{\partial x}\right) + \frac{\partial}{\partial y}\left(\epsilon \frac{\partial u}{\partial y}\right) = 0 \qquad (14.9.1)$$

$$u = g = \text{constant} \qquad \text{on } \partial G \qquad (14.9.2)$$

where $\epsilon = \epsilon(x, y)$ is the dielectric constant and is assumed to be known and ∂G is the boundary of G, that is, the surface of the material conducting the electric charge. The present problem is an *exterior problem* involving the exterior domain bounded by ∂G. Then

$$e = \frac{1}{4\pi} \int_{\partial G} \epsilon \frac{\partial u}{\partial n} \, d\sigma \qquad (14.9.3)$$

is the total electric charge on the conductive material, and

$$C = \frac{e}{g} = \frac{1}{4\pi g} \int_{\partial G} \epsilon \frac{\partial u}{\partial n} \, d\sigma \qquad (14.9.4)$$

is the electrostatic capacity of the conductive material located in the dielectric material. Here $\partial/\partial n$ represents the derivative toward the outward normal seen from the exterior domain G, that is, toward the inside of the conductive material.

Although there is some difficulty in solving an exterior problem by the FEM, we assume that a finite element solution can be obtained for this problem. It is easy to see that the given equation (14.9.1) is the Euler's equation of the functional

$$J[u] = \frac{1}{2} \iint_G \epsilon \left\{ \left(\frac{\partial u}{\partial x} \right)^2 + \left(\frac{\partial u}{\partial y} \right)^2 \right\} dx \, dy \qquad (14.9.5)$$

On the other hand, if we substitute the exact solution u^0 satisfying (14.9.1) and (14.9.2) into u in (14.9.5) and integrate by parts, we have

$$J[u^0] = -\frac{1}{2} \iint_G u^0 \, \text{div} \, (\epsilon \, \text{grad} \, u^0) \, dx \, dy + \frac{1}{2} \int_{\partial G} u^0 \epsilon \frac{\partial u^0}{\partial n} \, d\sigma$$

$$\qquad\qquad (14.9.6)$$

$$= \frac{1}{2} g \int_{\partial G} \epsilon \frac{\partial u^0}{\partial n} \, d\sigma = 2\pi g^2 C$$

This means that $J[u]$ corresponds to the physical quantity, that is, the *electrostatic capacity*. It is evident that $J[u]$ attains its minimum when $u = u^0$. On the other hand, the equation in the weak form corresponding to the minimization of (14.9.5) is given by

$$\iint_G \epsilon \left(\frac{\partial u}{\partial x} \frac{\partial v}{\partial x} + \frac{\partial u}{\partial y} \frac{\partial v}{\partial y} \right) dx \, dy = 0 \qquad \forall v \in \mathring{H}_1 \qquad (14.9.7)$$

$$u = g \qquad \text{on } \partial G \qquad (14.9.8)$$

hence, substituting a finite element solution $\hat{u}$ of this equation into (14.9.5), we obtain an approximate value C_1 of the electrostatic capacity C. This gives an upper bound of C.

In the present problem

$$F(u, u_x, u_y) = \tfrac{1}{2} (u_x^2 + u_y^2) \epsilon \qquad (14.9.9)$$

and F does not include u as its independent variable explicitly. Therefore, as in Sec. 14.5, we see that the corresponding variational problem is as follows.

Under the constraint that

$$\text{div } D = 0 \qquad D = (d_1, d_2) = (\epsilon \lambda_1, \epsilon \lambda_2) \qquad \text{in } G \qquad (14.9.10)$$

maximize

$$I[d_1, d_2] = -\frac{1}{2} \iint_G \frac{1}{\epsilon} (d_1^2 + d_2^2) \, dx \, dy + g \int_{\partial G} \left(d_1 \frac{\partial x}{\partial n} + d_2 \frac{\partial y}{\partial n} \right) d\sigma$$

$$\qquad\qquad (14.9.11)$$

The Euler's equation for this maximum problem is, from the final result in Sec. 14.5, as follows.

Under the constraint that

$$\text{div } D = 0 \qquad \text{in } G \qquad (14.9.12)$$

solve

$$\epsilon \frac{\partial u}{\partial x} = d_1 \qquad (14.9.13)$$

$$\epsilon \frac{\partial u}{\partial y} = d_2 \qquad (14.9.14)$$

$$u = g \qquad \text{on } \partial G \qquad (14.9.15)$$

The finite element solution $\hat{D} = (\hat{d}_1, \hat{d}_2)$ gives an approximate value C_2 which bounds C from below. Thus we obtain the following error estimate:

$$\frac{1}{2\pi g^2} I[\hat{d}_1, \hat{d}_2] \leq C \leq \frac{1}{2\pi g^2} J[\hat{u}] \qquad (14.9.16)$$

The exact solution u^0 of the original minimum variational problem that minimizes $J[u]$ under the constraint (14.9.2) gives the electrostatic potential. On the other hand, the exact solution D_0 of the dual variational problem that maximizes $I[d_1, d_2]$ under the constraint (14.9.10) gives the *electric flux density*. In other words, in the former problem involving minimization, (14.9.12) or, equivalently, (14.9.1) is solved to obtain the potential approximately under the condition that (14.9.13) through (14.9.15) are satisfied exactly, while in the latter problem involving maximization, (14.9.13) through (14.9.15) are solved to obtain the electric flux density approximately under the condition that (14.9.12) is satisfied exactly.

As seen above, for the dual variational principle the equations that the exact solution should satisfy are divided into two groups, and a solution is sought that approximately satisfies the equations belonging to the first group under the condition that it satisfies exactly the equations belonging to the second group. Then the physical quantity being considered, say the electrostatic capacity, is expressed in two different forms, each of which is a function of either of two different physical quantities, say the electrostatic potential and the electric flux density. An upper bound is then obtained from the approximation of one form, and a lower bound is obtained from the approximation of the other form.

14.10   Mixed Method

As seen from the discussion so far, the basic idea of the dual variational principle involves dealing with the constraint using the method of Lagrange multipliers. By extending this idea beyond the framework of minimization or maximization of the functional, we can apply it to a wider class of practical problems.

Consider Prob. 5 in Sec. 14.5 as an example. While the admissible function must satisfy

$$\text{div } \boldsymbol{\lambda} = -f \qquad \text{in } G \qquad (14.10.1)$$

in this problem it is generally impossible for a piecewise linear function to satisfy this condition exactly. We multiply (14.10.1) by $-u$ and add it to the functional $I[\lambda_1, \lambda_2]$ in (14.5.8):

$$I[\lambda_1, \lambda_2, u] \equiv -\frac{1}{2} \iint_G \boldsymbol{\lambda} \cdot \boldsymbol{\lambda} \, dx \, dy + \int_{\partial G} \left(\lambda_1 \frac{\partial x}{\partial n} + \lambda_2 \frac{\partial y}{\partial n} \right) g \, d\sigma$$
$$- \iint_G u \, (\text{div } \boldsymbol{\lambda} + f) \, dx \, dy$$

$$(14.10.2)$$

where $-u$ as multiplied above corresponds to the Lagrange multiplier. If we write down the condition that the first variation of $I[\lambda_1, \lambda_2, u]$ vanishes and write $\delta\lambda_1$, $\delta\lambda_2$, and δu as μ_1, μ_2, and v, respectively, we have the following system of equations in the weak form:

$$\iint_G (\boldsymbol{\lambda} \cdot \boldsymbol{\mu} + u \, \text{div } \boldsymbol{\mu}) \, dx \, dy - \int_{\partial G} \left(\mu_1 \frac{\partial x}{\partial n} + \mu_2 \frac{\partial y}{\partial n} \right) g \, d\sigma = 0$$

$$\forall \boldsymbol{\mu} = (\mu_1, \mu_2) \qquad \mu_1, \mu_2 \in H_1 \qquad (14.10.3)$$

$$\iint_G (\text{div } \boldsymbol{\lambda} + f) v \, dx \, dy = 0 \qquad \forall v \in H_1 \qquad (14.10.4)$$

Integration of the first equation by parts results in

$$\iint_G \left\{ \left(\lambda_1 - \frac{\partial u}{\partial x} \right) \mu_1 + \left(\lambda_2 - \frac{\partial u}{\partial y} \right) \mu_2 \right\} dx \, dy$$
$$+ \int_{\partial G} \left(\mu_1 \frac{\partial x}{\partial n} + \mu_2 \frac{\partial y}{\partial n} \right) (u - g) \, d\sigma = 0 \qquad (14.10.5)$$

Therefore, since μ_1, μ_2, and v can be taken arbitrarily, we see that u satisfies (14.10.3) and (14.10.4) satisfies (14.5.14) through (14.5.17). We chose $-u$ for the Lagrange multiplier because we expected this result beforehand.

The system of equations in the weak form (14.10.3) and (14.10.4) includes only the first-order derivatives, and there is no other constraint. Hence this system of equations can be solved easily by means of the FEM based on piecewise linear functions.

In this method we construct a functional by mixing two different physical quantities u and λ and obtain an approximate solution by making the functional stationary. This method is just the mixed method explained in Sec. 7.7. It is recommended that the mixed method using Lagrange multipliers be applied to a problem with constraints that are difficult to deal with directly. In the example shown above, however, the exact solution $(\lambda_1^0, \lambda_2^0, u^0)$ gives neither the minimum nor the maximum of $I[\lambda_1, \lambda_2, u]$ and is only a saddle point. Therefore, in this case, we cannot obtain an error estimate based on the minimum or maximum property of the functional.

References

1. G. Strang, G. J. Fix, *An Analysis of the Finite Element Method,* Englewood Cliffs, N.J.: Prentice-Hall, 1973.
2. K. J. Bathe, E. L. Wilson, *Numerical Methods in Finite Element Analysis,* Englewood Cliffs, N.J.: Prentice-Hall, 1976.
3. J. J. Connor, C. A. Brebbia, *Finite Element Techniques for Fluid Flow,* London: Butterworth, 1976.
4. P. G. Ciarlet, *The Finite Element Method for Elliptic Problems,* Amsterdam: North-Holland, 1978.

The purpose of this book is to describe the FEM as a technique for solving partial differential equations from the standpoint of applied mathematics, and [1] is the most well-established book written from the same standpoint. The author has referred to several chapters in [1]. Although [2] is written from the standpoint of structural mechanics, it acts as a bridge between use of the FEM in structural mechanics and in partial differential equations. Also, [2] includes a full description of the generalized eigenvalue problem $Ky = \lambda My$ appearing in Sec. 10.2. [3] is a text on the FEM as applied to fluid flows. [4] presents an FEM for elliptic problems and requires specialized mathematical knowledge of advanced analysis. The author suggests that readers who are interested in the mathematical aspect of the FEM study [4].

5. R. Courant, D. Hilbert, *Methods of Mathematical Physics,* Vols. 1 and 2, New York: Interscience, 1953.
6. T. Kato, *Henbunhou in Ouyousuugaku-Gairon, Ouyou-hen (Varia-*

tional Method in Applied Mathematics for Natural Scientists, in Japanese), Tokyo: Iwanami Shoten, 1960.

 7. G. D. Smith, *Numerical Solution of Partial Differential Equations,* New York: Oxford University Press, 1965.

 8. K. Yosida, *Functional Analysis,* Berlin: Springer-Verlag, 1965.

Although the titles listed above are not intended to describe the FEM, they are closely related to this book. [5] is a well-established book about the variational principle and partial differential equations. [6] can be referred to when studying the details of the dual variational principle in Chap. 14. [7] is a book on the finite difference method that is another powerful tool for the numerical solution of partial differential equations. [7] can be referred to when studying the time difference method discussed in Chaps. 8 to 12. [8] is a good book for those who want to study functional analysis.

9. G. E. Forsythe, C. B. Moler, *Computer Solution of Linear Algebraic Systems,* Englewood Cliffs, N.J.: Prentice-Hall, 1967.

10. A. Ralston, P. Rabinowitz, *A First Course in Numerical Analysis,* 2nd ed., New York: McGraw-Hill, 1978.

11. J. H. Wilkinson, C. Reinsch, *Linear Algebra: A Handbook for Automatic Computation,* Vol. II, New York: Springer, 1971.

12. D. S. Kershaw, The incomplete Cholesky-conjugate gradient method for the iterative solution of a system of linear equations, *J. Comput. Phys.,* 26:43–65, 1978.

The books and the paper listed above are on the solution of systems of linear equations and eigenvalue problems. [9] is a well-established book on the solution of systems of linear equations with a careful analysis of round-off errors. In [10] numerical analysis of linear algebra is given. [11] is a handbook of linear algebra computations with a variety of programs. In the first part of [2] there is an introduction to linear algebra. For information on the incomplete LU decomposition discussed in Sec. 4.11 readers should refer to [12].

13. H. Fugii, A note on finite element approximation of evolution equations, Mathematical Theory of the Finite Element Method, Kokyuroku No. 202, RIMS, Kyoto University, 96–117, 1974.

14. M. Tabata, A finite element approximation corresponding to the upwind finite element differencing, Memoires of Numerical Mathematics, No. 4, 47–63, 1977.

15. M. Mori, Stability and convergence of a finite element method for solving the Stefan problem, Publ. RIMS, Kyoto University, 12:539–563, 1976.

16. M. Hinata, M. Shimasaki, T. Kiyono, Numerical solution of

Plateau's problem by a finite element method, *Math. Comp.*, 28:45–60, 1974.

Recently, Japanese numerical analysts have contributed significantly to the development of FEM, particularly in relation to time-dependent problems, and a part of this book is devoted to some of these contributions. The stability of schemes discussed in Chaps. 8 and 9 is based on [13], the upwind approximation in Chap. 11 is based on [14], the application to the Stefan problem in Chap. 12 is based on [15], and the minimum surface problem in Chap. 13 is based on [16].

17. C. A. Brebbia, *The Boundary Element Method for Engineers,* New York: Halsted Press, 1978.

The boundary element method is well known as an alternative method for the numerical solution of partial differential equations. Since this subject is beyond the scope of this book, readers who are interested in this procedure should refer to [17].

Index

Index